LANGUAGE DISTRIBUTION IN DATABASES:

An Analysis and Evaluation

by

Gretchen Whitney

The Scarecrow Press, Inc.
Metuchen, N.J., & London
1990

British Library Cataloguing-in-Publication data available

Library of Congress Cataloging-in-Publication Data

Whitney, Gretchen, 1948-
Language distribution in databases : an analysis and evaluation / by Gretchen Whitney.
p. cm.
Includes bibliographical references.
ISBN 0-8108-2323-3
1. Foreign language publications--Bibliography--Databases--Evaluation. 2. Science--Bibliography--Data bases--Evaluation. 3. Bibliographical services--Evaluation. 4. Communication, International. I. Title.
Z699.5.F67W45 1990 90-34219
016.5---cd20

Manufactured in the United States of America

Printed on acid-free paper

CONTENTS

ACKNOWLEDGMENTS

This project could not have been completed without the able assistance of Ken Furuta, who gathered materials, commented on the text, and did his best to keep me organized during the process.

Charlie Bourne of DIALOG Information Services, who provided research funds for the study, was also essential to its completion.

I am grateful for those interviews which contributed to the understanding of the problems of database development: Karen Patrias, Senior Resource Specialist, National Library of Medicine, for the MEDLINE database; Clifford Backrach; Sally Lawler, Mathematical Reviews; Carolyn Gosling, PsycInfo; Gerard Plateau, Chemical Abstracts Service; Alice Picon, New York Public Library; Kathleen Deckerd and John Sears, Oceanic Abstracts; and Nancy Pearse, National Criminal Justice Reference Service.

Finally, my friends and colleagues provided emotional and intellectual support in the completion of the project, for which I am most grateful.

Gretchen Whitney
Graduate Library School
The University of Arizona

CHAPTER 1

INTRODUCTION

A common assumption and conclusion of many studies of the linguistic structure of scientific literatures is that English has become the language of science. While Latin and German once were the dominant languages for communication of scientific information, these languages largely have been replaced by English. We see the area of "English as a Second Language" becoming a defined field of study. We see students in Asia and Latin America flocking to English language classes. We see libraries increasingly dropping subscriptions to periodicals that are in languages other than English. We see scientific periodicals published in Europe adopting English titles rather than French or German ones.

Is this assumption correct? That is the subject of this book. There is conflicting evidence, which makes the question difficult to answer readily. There are numerous studies of specific bibliographies or bodies of literature that show that the use of English is indeed increasing. However, in magnetics, over the last twenty years, for example, Japanese papers presented at the Magnetic Society have increased to the point where they are the majority. Pharmaceutical researchers, for example, point out the need for results of experiments from countries with less stringent research guidelines. Databases are becoming readily available in the United States that are entirely in German or French.

To whom does it matter? Certainly young scientists in countries with national languages other than English are concerned about their ability to read and use the latest scientific writings, and they need to know what languages to study in order to both use scientific literature and contribute to it. Young scientists in English-speaking countries also need to know how to prepare themselves linguistically for their careers. National policy makers need a better understanding of the languages of science as they plan for national language policies. Librarians, both English-speaking and non-English speaking, often serve bi-lingual or multi-lingual populations, and generally collect in languages other than English. They need information about the languages of science and publishing generally to plan their collections, and to develop computerized systems able to handle these languages. The general area is also of concern to those interested in the continuing international interconnections that are developing as telecommunications and computers continue to shrink the size of the planet.

After the launch of Sputnik in 1957, the scientific community went into high gear to obtain Russian scientific and technical information. During this period, a variety of projects were initiated to acquire, translate, and disseminate Russian information. Language programs were started, and other projects were initiated to meet the scientific and technical challenge. Many areas in the scientific community are now experiencing a similar state of agitation over Japanese literature. This study shows that attention is now focused on both Chinese and Japanese literatures, and decreasingly on European work.

Not everyone is happy about the movement to English. There is concern for national pride and for cultural identity. Issues of language clearly have caused upheavals in Arizona, in Canada, and in Asia. Clearly, one wants to write in a language in which one is comfortable and able to express ideas; but one also wants to be read. Many developing countries see it as increased cultural domination by the west. Others view the study of languages as a humanistic, culturally broadening activity, and like Stavik (1987) assert that English-speaking countries also should resist this domination.

Science as an International Activity

Science by definition is an international activity: hence the concern for the language distribution of records in databases offered by secondary services.[1] It is through these secondary services, primarily electronic databases that can be searched from anywhere in the world with a computer or terminal and telephone line, that scientists find out about the international literature of interest in their fields. In an international activity, one would expect the appearance of materials in different languages.

Science is international in that "a single physical reality" exists which is to be investigated, and against which findings can be verified (Storer, 1970). As he notes, findings about X-rays will be the same no matter in what country they are discovered. Further, knowledge about this reality must grow systematically, based on previous findings. "If basic physical phenomena are universal ... it follows that the location of a scientist in space (if not in time) is irrelevant to his chance to aid in this advance--*provided he has adequate access to information on the current state of the art in his field* ..." (Storer, 1970). Access to recorded knowledge, through secondary services therefore becomes increasingly important.

We can expect variation by subject or discipline, however: not all sciences are international to the same degree. There are areas in science that are local in concern and in application, in particular in the social sciences. Studies of high density living in cities, for example, are of less concern in sparsely populated areas. Studies of desert wildlife are not of interest to researchers in the humid tropics. But this is more a condition of concern or application: studies of desert creatures add to the body of knowledge about the animal life of the planet, which is one of the objectives of biology.

---

[1]"Secondary services" are bibliographies which serve as indexes to primary scientific literature published in books, journals, conference proceedings and the like. They provide subject access to the literatures, and often include abstracts or summaries of the works indexed. In the late 1960's and 1970's most of them moved from printed to electronic form.

Rose and Rose (1969) suggest that while "the history of science is studded with examples of science as an international endeavor," internationalism in reality may be little more than a myth. Certainly some areas demand international cooperation to accomplish anything, such as studies of eclipses and other major natural events, or geographical and astronomical surveys. But even at the time of that writing, travel was becoming increasingly difficult for political reasons, and more and more defense-sponsored research was being conducted with restrictions on publication of results. They suggested that freedom to publish is restricted to a minority elite who are perceived as scientists, and it is these individuals who are making claims about the internationalism of science.

Storer suggests that the consequences of internationalism are a need for language abilities, a desire for quantification of results, and the migration of scientists to locations where science is being conducted--in the face of difficulties of adjustment and nationalistic pressures.

In part, the degree to which this "internationalism" appears is the subject of this work. It addresses the question of the English language dominance of databases, and the "ethnocentrism" or preference for English despite the international nature of the field in question.

Clearly, "internationalism" also manifests itself in trends in the incidence of materials *from* different countries. Where those data are available, they also have been examined.

United States Policy Positions Toward International Science

The United States government clearly sets policies which make it more or less possible for its scientists to participate in international science. Travel policies, financial commitments, and import/export regulations are a few examples of these.

Foreign policy, in part, can affect the availability of materials from other countries: in 1971, then Secretary of State William Rogers asserted that "It is the general policy of this administration to permit the exchange of unclassified scientific and technical information with the scientists of any country, regardless of the state of our diplomatic relations with that country" (Rogers, 1971). It is all well and good to permit the flow, but whether that policy is acted on, i.e., whether materials from other countries are indeed included in secondary services, is evident in this study.

Under the Reagan administration, the United States government took a variety of steps to prevent national scientific and technical information from falling into the hands of the Soviets (Boffey, 1982). In 1982, at a meeting of photo-optical engineers, highly sensitive information on satellite-to-submarine communication, was nearly given away, to the dismay of the government. The Department of Defense in that same year proposed a primarily military-composed committee to pre-censor scientific work supported by defense contracts and about to be made public (Strategy aims... 1982).

Even now, the Defense Department is struggling with two different approaches to the identification and use of foreign science and technology. The first is to actively seek it out and use it, to save development costs. The second is an international version of the "not invented here" syndrome. Foreign science and technology should be ignored and research should be conducted in this country, to prevent the possible release of sensitive information.

The National Science Foundation (NSF), long a supporter of international work, continues the first position. The National Science Board of the NSF, in a report that was to be issued in January of 1989, recommended that the Foundation "encourage international exchanges of scientific information" and "facilitate the translation and domestic distribution of foreign scientific literature" (McDonald, 1988).

The Developing Countries

While developed countries such as the United States have the resources to participate in international science, for countries without such resources the situation is more difficult. By and large, these countries (with the exception of India and some African countries) do not use English as a national language.

The developing countries are not late entries in science itself. Much of mathematics, for example, is historically traced to the Middle East and India. Generally, in contemporary science, however, they are not participants at the same level as industrialized countries such as Germany and France. The definition of "developing country" is elusive. While some countries like China might be considered less than industrialized economically, their science is becoming "world class." China may be the only example of this combination, however.

Goonatilike describes a structure of the movement of scientific information as having a center (the West) and various peripheral areas (the Third World). Dependence is caused by and reinforced by historical and economic factors. The center provides education and training for the periphery, and sets the research agenda and methodology. It also validates any discoveries made in the periphery. Only after this process are the findings then transferred through the center to another peripheral area. This process is dependent upon the literature in the center, and reputations are made by non-scientific (e.g., political) means. The development of English as a dominating language reinforces the idea of "science at the periphery."

Indian science serves as an example of post-colonial science. It is characterized by the remnants of a British educational system, geared not so much for research as for the production of trained students. Research is primarily done at the various research institutes. If scientists wish to be published, they read the Western journals, find out what is being worked on, and adopt similar work as their own. Informal communication is difficult between Indian and Western scientists, and virtually non-existent within the Indian community itself. Science comes by overseas mail (Goonatilike, 1984). While India can claim the third largest scientific community in the world (the first two being the United States and the Soviet Union, respectively), "qualitative" measures show it virtually unnoticed in the scientific literature. Publications cite older materials and other Indian sources. They are not cross-disciplinary. And India is considered a "leading scientific power in the Third World."

The developing countries are in a double-bind, because the level of scientific activity which they would like to achieve requires a level of participation in the international community which for economic and social reasons is denied them. At the second (1966) meeting of the science ministers of the Organization for Economic Cooperation and Development (OECD), they "recognized the need for each country to sustain sufficient research activity to provide a threshold of awareness of world scientific progress as well as of advanced skills, without which the transfer of technology is unlikely to be effective." It is this threshold which serves as the platform from which they can participate, and it requires at the very least a recognition (from other countries) that it has been achieved. This recognition virtually demands a bi-directional flow of information, each side receiving something of perceived value. Even in information, countries must have something of value to trade.

The issue of language can be linked even more deeply in developing countries to education and textbooks. De Silva describes the dependence of Sri Lanka on English language materials for education at all levels, but especially for higher education. British texts are retained as teaching materials and are kept in "prized collections" in libraries. He charges that "the language barrier ... is at the bottom of the failure to de-colonize our society."

The World's Languages

One of the most interesting questions for background consideration is the number of languages actually spoken in the world, and the number available for contribution to scientific and technical literatures. As Shearer (1985) notes, it is virtually impossible to determine the number of distinct languages in use. He cites estimates of between 3,000 and 5,000 languages, exclusive of dialects. He suggests that there are 197 languages spoken by at least a million people. And, half of the world's population uses one of eight languages as their mother language: Mandarin (China), English, Russian, Spanish, Hindi, Arabic, Bengali, or Portuguese.

Notice that this may be the last time we discuss the Bengali, Arabic, and Hindi languages. They are not languages that appear frequently in discussions of the languages of science. Notice also the absence of Japanese, German, and French. These rank 9th, 10th and 11th respectively in populations of native speakers by languages spoken worldwide (World Almanac, 1988).

The Growth of English

Why did English become such a dominant language? It was not expected to do so in the 1600's, being rivalled by French, Spanish, and German (Bailey and Görlach, 1982). John Milton, writing in the middle 1600's, could have as easily (by his contemporaries' expectations) written in Latin as in English. In the same century, Francis Bacon wrote in both English and Latin to ensure that his scientific works would be read (Russ, 1982). In fact, there was a "strong... resistance to the use of English for the most serious purposes of life" due to its instability (Russ, 1982).

Between 1800 and the middle of the 20th century, the "number of native speakers of Spanish has doubled, German trebled, and Russian multiplied by five, [and] English ... expanded by a factor of ten" (Starr, 1978). He attributes the spread of English primarily to the expansion of the British Empire, but also notes the development of North America and Australia, and lengthened life expectancies in English-speaking countries through rapidly advancing medical science and technology, without a decrease in the birth rate.

Bailey and Görlach (1982) compare the spread of English to that of French during the colonial period of France. It declined in prevalence as the colonies were surrendered. They also note that when English is brought to a country, it remains and is not replaced by another language. Other languages implanted during colonial times have been often eradicated (such as French and Spanish to a large degree from the former colonies in the United States).

Starr (1978) suggests that English has overturned French in many circumstances. It is studied in preference to French, which had been the favored second language in Brazil, Colombia, and Venezuela. In France, English is studied rather than German, Spanish, or Italian.

The fact that a country has designated English as its national language does not necessarily mean that it is spoken widely throughout the country. As Todd (1982) points out, the varieties of English in West Africa are numerous, and range from a "local standard" through a "pidgin/Creole-influenced English" to a "very basic sort of English consisting primarily of words ('bring money') strung together." Granted, a university-trained student will have stronger skills than an uneducated person working in the local market. But, the members of certain less educated groups, who need English in their work, may have, as Todd points out, stronger skills than a student with several years of primary education.

Obviously, a wide range of diversity in language abilities is also present in the United States and Great Britain. Literacy rates, however, suggest that the range may be more severe in the West African countries. The point is that just because a country's official language is English, we cannot assume equal linguistic skills. And we must recognize, as Todd (1982) suggests, that these various levels interact in "education, commerce, trade, bureaucracy, in the science and literature ... [to] permit communication and interaction."

East African traditions are different from West African, because European settlements in East Africa were much stronger, more permanent, and influential. West African Pidgin English, which developed as a secondary language, never materialized in the East. But, English infused into the region through trade, through British missionary workers and non-British workers who studied English before going to Africa, and through the educational systems that they established. It was also brought in by slavery in the 1400's and 1500's. It was "common" by 1800 (Todd, 1982). Hancock and Angogo (1982) conclude that English will remain in East Africa. It was declared the "lingua franca" in 1929. Local languages predominate in Ethiopia, Somalia and Tanzania, however. Where it remains, it does so in part because of its "prestige value," but also because advanced educational materials are available only in English.

Comparing French to English, Bailey and Görlach (1982) suggest that the insistence by the French government on a single, standard, impeccable language has somehow given a certain mystique to the language, which makes it less attractive for adoption. English, with a multitude of de facto standards, is "emotionally more neutral," contends Ali A. Mazrui, an Ugandan scholar whom Bailey and Görlach quote, and this fact more readily permitted the development of a "national consciousness" through English.

Will English maintain its dominance? Starr (1978) thinks not. He predicts the rise of developing country languages to take their place in a more multilingual world. He notes Tanzania's change to Swahili, India's commitment to de-Anglicization for official business, and the general trend for former British colonies to adopt local languages for education. As noted above, he points to declines in English-language publishing in Asia and Africa as symptomatic of this trend. He also notes the rise of research activities in other countries, and it was the rise of research in the United States that gave English momentum in science and technology. We must note that just because a country is increasing its R&D does not mean that the results of that work will be published in the local language: rather, since those scientists want to be heard, it is more likely that they will publish in English and increase the total English language production.

He also indicates the fragility of the dominance of the language in certain circumstances, such as dependence on other countries for resources. If, for example, "Saudi Arabia decide[d] tomorrow that it would prefer to conduct all its negotiations on oil prices in Arabic, the United States would scarcely be in a position to refuse."

National Language Planning

Language planning is a part of the entire social and economic planning process, argues Zakhariev (1980). In the case of multilingual developing countries, these decisions are most difficult and complex, and involve issues of development of a leadership cadre in all areas of the economy, while at the same time maintaining the use of the mother tongue to keep the cultural heritage. They also involve issues of national unity, which can be torn between the native language and one imposed for years by a foreign power: both of which suggest unity. He argues that no one language is inherently less able to handle the concepts of science and technology, and that the scientists of a given language should develop the necessary vocabulary needed to "modernize" the language. That indeed may be a lofty goal, but who are such scientists going to talk to, other than each other? That, for reasons political and psychological, may be deemed to be important. Is the development of a scientific vocabulary in a local language a necessary expenditure of scarce scientific and technological resources? For political, cultural, and social reasons, it may well be so. We should note that the author is Bulgarian, and he points to the multilingual USSR as disproving the assertion that a multi-lingual country cannot achieve a high level of scientific and technical development.

Kaplan (1983) summarizes the difficult options facing new countries as they choose a national language. They must choose one of those which enables them to have access to world science and technology (English, French, German, or Russian), over one to be chosen for "cultural or religious solidarity" or the one spoken by the largest number of its population. He suggests that despite the political attractiveness of other decisions, use of a world language will improve their chances for development. He barely stops short of recommending English as the language of choice.

The Problems of Non-English Literatures

The appearance of materials in languages other than the scientist's own, and the inaccessibility of texts in foreign languages, has been cited as a problem in a variety of sources (Chan, 1976; Anderson, 1974; Ellen, 1974 for example). Most of the studies offer suggestions for improvement such as translation services, or training in English as a second language for young or established scientists. Generally the studies are "snapshots" of the language situation at a given point in time, and do not indicate whether the problem is getting any better, or worse, and for whom.

Shearer (1985) asserts that "... it is the language of the text rather than the nationality of the author that is frequently most important under prevailing international communication systems ..." In terms of access to literatures, whether scientific or not, he is precisely right. While governments can regulate the import of materials, generally they regulate by publishing nation or by subject, not language. Advancements in science are generally discussed in terms of the accomplishments of a nation and the specific area of research, such as "French computer technology." One does not speak generally of advances in French scientific literature per se. But when it comes to the individual scientist, given the relatively free flow of scientific literatures across national boundaries, it is the language of the text, rather than the country of publication, that is the final determining factor in whether a document can be used.

Shearer places the study of languages themselves and languages as represented in different literatures in a broader context of understanding the "macro concerns" of a global information system: the information that describes "its users and potential users and their potential uses for it," as opposed to "micro" concerns of the information that is contained in the system itself. These larger concerns generally are planning concerns. One assumes that the global information system of which he speaks is not a single system, but the total available set of information systems, interconnected. He terms one aspect of the planning agenda, "geolinguistic ecology," or where and how a system's information is to be made available to people. It includes "the interrelations of the information skills and behaviors of people, the properties of the information system's content, and the location of its service points." He suggests that a critical problem is that we just do not have solid information on who has access to what materials in what languages.

The Significance of the Language Problem in Databases

The trends in the language distribution of materials in databases can have an impact in several areas, such as policy planning for languages and education. For the individual scientist, however, for whom these databases largely were developed, the problem is access to the work of colleagues around the world. The secondary services lay out for the scientist the literature available in his or her language: if the scientist's first language is English, he or she may be being denied access to important works of colleagues in other languages. If the scientist's native language is not English, access may be denied to works in his or her own language because records for the works are excluded from the database, and the scientist may be forced to handle works in one in which are less comfortable.

This bias toward English is significant in two areas: the degree to which scientists themselves are comfortable in the languages represented, both through birth and through formal training, and their nation's overall contribution to world science and technology. Regarding the first, if a great deal of literature is being published in a language that scientists cannot read, then its usefulness is diminished for those scientists. We must assume that all scientists cannot read all of the languages in which science is published.

The increased use of English means that scientists who are not native speakers of English must learn English or obtain help in other ways to use the literature. Regarding the second, if a nation or language group is not active in an area of science, literature is not expected to appear in that language. That there is not a tremendous literature from scientists writing in tribal languages is not of serious concern. There are not very many scientists speaking those languages (in relation to the world's languages) nor writing in it: nor could they expect many scientists outside the country to be able to read reports in those languages. (It can be acknowledged wholeheartedly that writings in tribal languages do have significance and cultural importance, however.) And, one typically does not look to developing countries for advances in, for example, aerospace engineering. If, however, there is a small representation in bibliographic databases of materials in French, but there is a significant number of scientists speaking that language and working in that area, then there may be a significant problem.

A predisposition to English in databases can come from several sources, the most significant being the agencies which produce databases themselves. (These issues are explored in greater detail later.) For United States agencies, there may be a lack of resources (financial and staff) to be able to handle non-English materials. The group developing the database may have a specific policy about their inclusion: indeed, non-English materials are more difficult and expensive to handle than English materials, and they may be omitted or set aside for later inclusion.

Some database providers are under pressure from a variety of sources to remove foreign language materials from their files. Sometimes the pressure is economic: indeed it is expensive to identify, acquire, and process non-English materials. Sometimes the pressure is from users, who see the foreign language materials as clutter.

Examining Languages in Databases: An Overview of This Monograph

The problem can be broken into two parts: first, in what languages is scientific literature being produced, and second, what is the nature and effectiveness of the access to that literature? Clearly, if there is a great deal of literature on, say, semiconductor technology being produced in German, but if no one includes it in a database, then no one will know about it and it will not be read (or its use will be greatly delayed.) The focus of this work is on the second question, but it cannot be understood without reference to the first.

This monograph is not the only work concerned with languages and access to literatures. Others have taken quite different approaches. For example, it is possible to look at citations used in works, to look at individual databases, and to compare databases on the same subject area. Many have examined the Institute for Scientific Information's Science Citation Index for indicators. These studies are the subject of Chapter 2.

We can take a broader perspective and look beyond single databases and look at the entire range of databases available. The growth of databases, their availability, and the languages covered by them shows the increasing internationality of these files. This growth and coverage is the subject of Chapter 3.

For this study, eight internationally oriented databases which purported to cover the world's literature in their discipline were chosen and examined in detail to determine their linguistic trends. These were BIOSIS, Chemical Abstracts, GeoRef, MEDLINE, NCJRS (Criminal Justice), Oceanic Abstracts, PsycInfo, and the Public Affairs Information Service (PAIS). The results of these findings are presented in Chapter 4. The Appendices provide detailed charts of the languages in selected databases, as described in this chapter.

It is the contention of this study that the policies and practices of the database provider and its parent organization have a strong influence in determining the degree to which the database includes non-English materials. The policies of the eight databases, and others, are discussed in Chapter 5.

As noted above, it is important also to compare the coverage presented by the secondary information services with figures for the actual literature produced. Databases by their nature are selective in what they choose to include: they cannot include everything. Raw production statistics by language are extremely difficult to locate, much less to evaluate and manipulate in a comparable form. The Unesco book production statistics published in the Unesco Statistical Yearbook are a frequently-cited source. These figures and others are examined in Chapter 6.

Chapter 7 summarizes the findings of the study and compares these new data to those presented from other sources.

CHAPTER 2

APPROACHES TO THE PROBLEM

Interest in the languages of science over the last three decades has been expressed by database producers, applied linguists, bibliometrics researchers, and librarians. Studies have included analyses of databases and bodies of literature showing their language content, use studies of the proportion of foreign language material, studies of the productivity and publishing patterns of specific countries, and assessments of how scientific information is communicated, that is, publication patterns and studies of formal and non-formal communication links.

The first group of studies are directly relevant, in that they represent other attempts to assess the proportion of different languages in databases. The second are peripheral but of interest, in that they represent assessments of the demand for foreign language materials in a variety of settings. The third are peripheral in that they are limited geographically. The fourth represent important supportive evidence of the way scientific information is communicated, which serves as the foundation on which secondary services are based.

Analyses of Databases and Bibliographies

Studies of databases and their structure often show a clear concern for the languages contained therein. The critical issue with them, however, is the degree to which the databases can be said to accurately reflect the literature they attempt to cover. It is common for the researcher to generalize from database findings to disciplines. For a variety of reasons that will be demonstrated, this is not always possible or wise. Nonetheless, these studies present some data for comparison.

Working with electronic databases is a fruitful approach, because of the structure of records in many files. Working with printed sources is tedious and time-consuming. On the other hand, online (or electronic) databases can be readily searched for specific data in specific fields. Most vendors have applied some degree of consistency in the way that records are laid out. And, in contrast to the time it would take to search printed services, since they can be searched with relative ease, a more comprehensive sample can be taken over a much wider number of databases.

The Rosenberg Study

An early attempt at analysis of multiple electronic databases for comparative purposes was Victor Rosenberg's initiation of a study of the foreign language content of about thirty-five databases in 1981. This was about the first time that such a comprehensive study could be attempted. The online vendors at this point had approximately ten years of data available for searching, which was sufficient to make possible a longitudinal study over multiple files in a reasonable time frame. The project showed the difficulties inherent in a project of this type. No report was ever issued on the data or findings.

He attempted to solicit language distribution data from selected database providers for fifteen specific languages for the years from 1969 to 1974. The providers largely were unhelpful, primarily because of the costs and time it would have taken to provide the information. (They would have had to conduct searches on their databases on DIALOG or another database provider if the data was not readily available from local files, which it frequently was not.) He also suffered many delays in responses because of staff changes at the database provider offices. And, since definitions of languages were not common across the board, the data were often not comparable. Of the responses received, six generated some useful information. Four of these are presented in Appendix 1. The American Chemical Society and the Institute for Scientific Information provided more extensive data, and these tables are also presented in Appendix 1.

The RILM (Répertoire International de Littérature Musicale) database statistics are included in Appendix 1.2. The data showed that for music, German by far was the most important language. This is not surprising, because of the Germanic origins of musicology in this country. Of the non-English records, 61% of them were German, followed by French, Russian, and Italian, each at less than 10% of the non-English records. Unfortunately, the totals for English materials were not provided, therefore language proportions cannot be determined. This was one of the few instances where a non-English language so clearly dominated the field, at least for these years.

METADEX (Metals Abstracts Index) from the American Society for Metals provided gross data for records entered from 1969-79. These were counts of records input in the years in question, rather than by date of publication. The database provider did not give the number of records in each language in comparison with the total number of records input for each period. The highest number of records were in Russian (86,702), German (26,429), French (13,078), Japanese (9,103) and Czech (4,018) for the entire file, apparently as of 1981. These represent about 89% of the file as reported. No yearly statistics were provided, nor were data for English records. The high incidence of Czech is interesting.

Information Retrieval Limited (IRL) provided data from the IRL Life Sciences Collection. The statistics for each year were by update year, rather than publication year, and did not give English language numbers so that proportions of the whole could have been determined. The data covered only a three-year period (1978-80). The most frequently cited languages were German, French, Russian, and Italian, with Russian showing a persistent increase over the three years. The first three languages made up 77% of all non-English records presented. These statistics are presented in

Appendix 1.1.

The American Geological Institute, providers of GeoRef (the Geological Reference File), supplied data only for 1970 and 1971. (This file is more completely evaluated in Chapter 4.) As presented, the data equated a language with a country: France was equivalent to French, for example, which is not accurate. Even for two years, the data showed a decline in French and German, and a doubling of Russian. These data were questionable for the present study, because they were not compiled in a way remotely resembling the methods used by other respondents.

The Chemical Abstracts data in Appendix 1.3 showed a high proportion of Russian materials in the set of non-English records, followed by German, Japanese, and French. Notice that these must be removed entirely from the chart for most other languages to appear at all. Of interest is the steady number of Bulgarian, Czech and German documents, the steady rise of Chinese, Japanese, Portuguese, Russian and Serbo-Croatian materials, and the slow decline of French. The Scandinavian languages showed an erratic pattern.

The Institute for Scientific Information (ISI) data from the Science Citation Index (SCI) for 1972-79 for selected languages (Appendix 1.4) showed a high proportion of German and French materials, and like Chemical Abstracts, these languages have to be removed to allow other languages to appear. SCI also showed a slow increase in Chinese, Danish, Japanese, Spanish and Swedish, and a more rapid rise in German and Russian. Italian documents declined and then rose for ISI, while in Chemical Abstracts they did the reverse: they rose, and then declined in number. Data for English-language records and records in other languages were not given, so the data could be used only for understanding gross trends in specific languages.

While the Rosenberg study began to take steps in the right direction, it demonstrated the difficulties in trying to develop consistent data sets for analysis when one is subject to someone else's interpretation of the problem at hand. Having greater control over data acquisition can lead to much more comparable data. Nonetheless, this project served as a stepping stone for the current inquiry.

Discipline Studies

Social, Natural, and Medical Sciences

Admitting the difficulties of defining what is "social science," Line and Roberts (1976) attempted to assess the literature as of 1973. They were unable to determine a language distribution, indicating that one source estimated that English contributions were 80% of the literature of the field, another estimated that it was 40%.

By subject, more European languages were found in journals in the natural and medical sciences. Engineering and agriculture ranked third and fourth as users of European languages, and the lowest ranking was given to chemical industries, photography, metallurgy as a group, and domestic science, business management and other subjects also as a group. This tends to support the idea that the more nationally applicable fields are discussed in the local language, and subjects applicable worldwide are more likely to be discussed in other languages to reach scientists worldwide.

Astronomy

In 1967, Meadows studied the citations referenced in the 1963-65 issues of the *Monthly Notices of the Royal Astronomical Society,* and found that English-language journals represented about "90% of the total references." German and Russian materials were cited less than 5% of the time, and French materials were cited about 2% of the time. This is somewhat surprising, because astronomy could be expected to be one of the more international fields.

Chemistry

The Chemical Abstracts database of the American Chemical Society consists of abstracts of documents from over 12,000 periodicals from 150 countries. Materials include patents, "as well as proceedings of symposia, dissertations, reports and books." The entire database, according to their staff, represents more than fifty languages. (U.S. House. Availability... 1984) It has been the subject of several analyses.

The Wood study reported that in 1965, Chemical Abstracts was 50.3% English, 23.4% Russian, 7.3% French, 6.4% German, 3.6% Japanese, 0.5% Chinese, and 8.5% Other languages.

Chemical Abstracts appears to be consistently concerned about its files, and reports regularly about changes in its contents. In Baker's 1976 report, he noted that Chemical Abstracts has retained the same "coverage policies" for the last 25 years, implying that changes in the database were the result of changes in the literature itself, rather than in any policies or practices at the organization. He did not mention any staffing changes, for example, which might have affected the ability of the service to include materials in different languages.

In that report he noted an increase in the proportion of English language materials from 1961-1976. In 1961, 43% of the file was in English. By 1965 the proportion had risen to 53%, by 1970 to 56.4%, and by 1975, it was 59.7%. This is consistent with Wood's analysis.

Baker said further that "The share of the literature originating in English-speaking nations has remained at about 40% throughout the period."

Regarding materials in Russian, he reported that the proportion from the USSR were what one would expect, when the few papers published in the USSR in languages other than Russian were taken into account. Russian documents consistently increased from 18.4% in 1961 to 23.3% in 1975.

Regarding German, French, Japanese and Italian, he saw a general decline in use between 1970 and 1975.

Chemical Abstracts includes an author's address field in its records, which can be analyzed to show the levels of activity in a given country. Note that it is possible for a scientist to work in a country that is not his own: therefore the figures do not, in a sense, represent the total activity that a country might *claim* as its own. Nonetheless, he made the following points:

The U.S. output increased in quantity between 1970 and 1975, but it declined in relative proportion to the total (from 27.4% in 1970 to 25.8% in 1975). In 1951, the contribution was 36.6%.

The percentage of contributions from the USSR increased slightly in the 1970's, from 23.6% in 1970 to 24.6% in 1975. Its contribution was 6.3% in 1951 and 19.1% in 1960, so its major period of growth was in the 1950's. Japan showed a decline over the same period. Canada increased sufficiently to rank seventh with India.

Fisheries

Because fish and ocean life know no national boundaries, and fishing certainly is of concern to any country with a coastline, one might expect to find a high proportion of non-English materials in the literature. Baldauf and Jernudd (1983) found, however, that English predominated in the 1978-81 files of Aquatic Sciences and Fisheries Abstracts with 78% of the records. Just 6.1% were in French, 3.7% were in Russian, 2.5% in German, and 1.9% in Spanish.

The authors further examined the file and found that the English materials were added to the file more quickly. This could be attributed to late arrival of documents due to postal delays. They also found that the more "statistical-type" reports were in English, while the non-English ones dealt more with the availability of different types of fish.

Abstracts are often intended to aid readers who do not speak the language of the text of the article, but this is not the case for this file. The abstracts generally were in the same language as the article.

The researchers also identified a sizeable number of articles written in English by authors from non-English countries. They interpreted this as an attempt on the part of the authors and journals to reach a broader audience.

Geography

In what one might expect to be a very international field, geography, Aiyepeku (1973) found that 77.4% of the works cited in *Current Geographical Publications* were in English. Only 8.8% were in French, and German accounted for but 8.2%. He determined that the language of the author's country generally determined the language used for publication.

History of Health Sciences

A 1982 study of the HISTLINE (History of Medicine) file on the National Library of Medicine's MEDLARS system revealed a diverse collection of records in other languages (Kopp, 1985). The United States contributed but 15.5% of the file and the United Kingdom 9.4%, other countries contributed the rest. The most frequently cited journal is a monthly Russian public health journal, followed by a Hungarian medical weekly. Of the ten most frequently cited journals, six are not in English. Of the twenty-two most frequently cited journals, fifteen are not in English.

Library and Information Science

Buckley contended that the coverage of foreign and foreign language materials in library and information science was increasing. *Library Literature* increased its foreign title coverage from four to six titles between 1979/80, *Library and Information Science Abstracts (LISA)* increased from nineteen to twenty-nine titles, and *Referativnyi Zhurnal (Section 59, Informatica)* increased from one to four. She attributed the *LISA* coverage of non-English materials to its market and subscribers, 25-30% of whom are in the Third World. A two-year study is a bit thin for proclamations of increased internationalism, but the study shows concern for the problem.

Physics

The Burke and de Sola Price study (1981) of the SCI database for physics focused on the degree to which different countries were international or insular. They found that advanced countries such as the United States, Soviet Union and Japan were insular, but countries such as West Germany and France were more international. The Frame and Carpenter study also pointed to insularity in these countries by studying co-authorship.

This is in contrast to the Pravdic and Aganovic-Boras (1988) study, which suggested the reverse: that the middle-level countries and those on the periphery were more insular than those less developed.

Country and Area Studies

Australian Research Literature

If works are not included in a database, it is as if they are not published at all. This was the concern of Byrne's study of Australian science. He reviewed coverage of several hundred Australian science and technology and social science journals in seventeen major files such as BIOSIS and MEDLINE, and found that in the best cases, only a third of the titles were given coverage, and in most cases it was significantly less. At least during the early 1980's, it was necessary to go beyond the major indexing and abstracting services for identification of Australian work.

Developing Country Literature

It has been suggested elsewhere that developing countries largely work on the periphery of science, and look to the more advanced countries for research ideas. Swales (1985) added to the discussion of this "imbalance" in a study primarily focused on authorship. In so doing, he also found that developing country literature simply has been being ignored by abstracting and indexing services. He does not, however, give an estimate of just how much literature that might entail.

Delays in appearance in databases of materials from developing countries is mentioned elsewhere in this chapter, and it is repeated by Christovão (1985) in his study of biomedical literature produced by developed and developing countries.

Guha (1985) attempted to explain the problems of developing countries, especially the "evil effects of excessive dependence on a foreign language." One of his primary concerns was that if developing country scientists learn and work in another language, they will not be able to communicate with the rest of the population of their country. His recommendations to resolve the problem reflected concern for translations, language education for librarians, and more studies of the problem. Like other studies concerning the developing countries, the author gives no indication of the extent of literature that is being missed by abstracting and indexing services.

French Scientific Literature

Eugene Garfield (1980) took French scientists to task for complaining about being ignored in the scientific community, and showed that according to the Science Citation Index, French is no longer a language of science, and French literature is only cited by the French themselves.

German Science Journals

While German science was pre-eminent before the second war, and German fluency was a must in many scientific disciplines, more recently we are seeing an Anglicization of their work. In the instruction for authors submitting work to *Naturwissenschaften,* they were directed: "English now is the Lingua Franca of the sciences" therefore manuscripts are to be submitted in that language. Several German journal titles have changed from German to English. Stankus, Schlessinger and Schlessinger (1981) examined a set of ISI German titles, and found "striking" movements to publish English articles. Swales (1985) also found such changes to German titles.

Japanese periodicals

The Foo-Kune study reported that the percentage of Japanese journals providing European language abstracts fell from 33.2% in 1957 to 17.9% in 1967, and the production of Japanese journals entirely in Japanese grew from 48.5% to 70.4% of all journals produced during the same time period. The Japanese literature did not become more manageable for the non-reader of Japanese. The authors estimated that the language barrier for these materials increased at about 2.2% each year.

Japanese Technical Literature

In 1985, the National Bureau of Standards and the Institute of Electrical and Electronics Engineers sponsored a conference on access to Japanese technical information, in particular engineering information (Brady 1986). The conference addressed the imbalance between the U.S. and Japanese engineering fields, in that the Japanese scientists can read U.S. technical literature, but the U.S. engineer cannot read that from Japan in Japanese. Brady, in his foreword, suggested that 20% of Japanese engineering is published in English, and wondered about the significance of that literature.

Ambler, in his charge to the seminar, stated that the reason for the interest in Japanese literature is economic: that because of the increasing Japanese technological presence in the U.S. marketplace. Japan will be increasing the competition: Japanese governmental statements through its five-year plan that indicate that Japan intends to become more "creative" by increasing basic science research, and intends to develop advanced technologies such as Fifth Generation computers. It behooves the United States technical community to pay attention to Japanese technical literature.

Latin American Social Sciences and Humanities

In contrast to studies indicating that the incidence of developing country work is declining in the secondary services, Vessuri asserts that it is in fact on the increase. Science and technology in Latin America is growing as indicated by funding as well as the formulation of science and technology councils and science centers. One would expect that this activity would result in increased publications as a result of the research, and indeed he cites three new serial titles in Spanish and Portuguese.

In contrast, Sandoval (1974) decried the fact that much Latin American work is being published outside the region. He examined *Science Citation Index* data for the late 1960's and demonstrated that his assertion was true. He repeated similar statements at a 1983 conference on North and South American information transfer (Rosenberg and Whitney, 1986).

According to Colson, Latin American materials can be monitored or found effectively on the databases in BRS, DIALOG, Wilsonline, the Mead files, NewsNet, and Questel. However, most of the records found were in English, and he thus concluded that for foreign-language materials, coverage was "barely adequate to very poor." He indicated that the *Hispanic American Periodicals Index* had not been successfully mounted on any of the vendor systems. He did express optimism for a CD-ROM version of the file.

Polish Scientific Literature

Stefaniak looked at the coverage of Polish literature in *Library and Information Science Abstracts (LISA), Chemical Abstracts, INSPEC,* the ISI *Science Citation Index,* and *Information Science Abstracts.* She found that coverage had decreased between 1977 and 1983 in *LISA,* from eighty-five papers to four. Decreases were also noted in the other sources. She found that Polish literature was much more strongly represented in *Referativnyj Zhurnal, (Section 59, Informatika)* at 3.5% of the file.

In Chemical Abstracts, she found that the applied chemistry titles were published in local journals, but the physical chemistry titles, more globally oriented, were published in the international journals.

Polish science, it appears, is suffering much the same fate as Australian science.

Soviet English Language Periodicals

Gloria Jacob's review of the coverage of these materials in current United States indexing services showed that the coverage is virtually non-existent. She attributed this problem to database policy: ISI did not view them as scholarly, and Wilson titles are selected by subscriber vote.

Soviet Psychology

It appears that the Soviets may have had as much difficulty keeping up with the United States as researchers in the United States have had in keeping up with their Soviet counterparts. In a study of Soviet articles on paranoid disorder included in the 1974-1983 issues of *Index Medicus,* Swanson and Swanson found that 41% of the articles included citations to United States research, but the citations were ten or more years old.

Spanish (Spain) Scientific Literature

Spanish publishing in science appears to support the idea that international fields are reported in English, whereas more local-interest work is published in the language of the country of origin. Studying eight of the major scientific databases such as Chemical Abstracts, Excerpta Medica, and BIOSIS, Mendez (1986) determined that the vast majority of Spanish scientists were publishing in foreign journals when they were identified by Chemical Abstracts, BIOSIS, and COMPENDEX (62-96%), but when they appeared in CAB Abstracts (Commonwealth Agricultural Bureaux) they were publishing in journals from Spain. The years studied were 1978 and 1982.

The authors determined that Spanish work was well represented in Excerpta Medica and INSPEC, because most scientists published in foreign journals. Pharmaceutical work was a problem in that only 50% of the most important journals were covered. Chemistry, agriculture, and engineering also were not well covered.

Despite this the authors concluded that more and more Spanish scientists were publishing internationally and in English, and that this was providing good visibility for their work.

World Technical Reports

The U. S. National Technical Information Service (NTIS) efforts have focused on the acquisition of technical reports produced primarily in private industry and by the government, and not on the open journal literature. In 1984, Shonyo reported that NTIS was acquiring about 800 documents a year from Japan, representing about 5% of their total foreign acquisitions (U.S. House... Availability... 1984).

Science Citation Index: Citation Studies

The Science Citation Index (SCI) database from the Institute for Scientific Information is based on citation indexing. It lists articles in selected journals, and the references included at the end of those articles. The underlying philosophy is that there is a subject or methodological relationship between a research paper and the papers it cites. While there are obviously a wide variety of reasons why one author will cite another, the database nonetheless has proved popular and useful. The SCI database has proved to be one of the most heavily analyzed files for language studies (Carpenter and Narin, 1977; Pravdic, Aganovic-Boras, and Kritovac, 1988 for example).

Computer Horizons, Inc., for the National Science Foundation, evaluated tapes of the SCI, and determined that "In 1982, U.S. scientist(s) and engineers were responsible for 35 percent of the world's influential S&T [scientific and technical] literature..." They considered this substantial, because "only 15 percent of researchers in industrial nations reside in the United States" (National Science Foundation, 1985).

The study did not detail exactly how "U.S." was determined, or give any other details. It also did not indicate what countries rose in percentage, in relation to the United States.

Given the frequency with which SCI data are analyzed for languages and country of publication it is not surprising that there have been investigations into its validity for the purpose. Carpenter and Narin (1981) worked with data for a 1972 study on various scientific disciplines, and compared their data with those in relevant subject databases. For mathematics they found that the United States, United Kingdom, West Germany and France were "overrepresented" in the SCI file, while the Soviet Union was "underrepresented." The SCI file contained significantly more Soviet work in Chemistry than Chemical Abstracts, but fewer Soviet works in biology than BIOSIS. SCI and Physics Abstracts were relatively in agreement for physics. For Soviet medicine,

MEDLINE was found to be the stronger file. For Soviet mathematics, Mathematical Reviews was found to be stronger.

They also compared SCI coverage with the holdings of the British Library Lending Division (BLLD). The former's collection policy is to seek "influential" work, while the latter is to collect anything with scientific merit. The BLLD collection showed 20% of titles with a United States origin, the SCI database included 39%.

This disparity was more dramatically clear for subject areas. For clinical medicine titles from East and West Germany, for example, the BLLD covered 399 titles and SCI covered 76. For the German engineering and technology, the BLLD covered 1346 titles and SCI 96. For biology titles from the United Kingdom, BLLD held 602 while SCI covered 508. Going further to test the usefulness of the BLLD data, which could readily have influenced the United Kingdom data for example, the authors compared BLLD data with *Ulrich's International Periodicals Directory* and found the two comparable.

Despite all of these potential reservations, the authors decided that SCI was appropriate for international studies.

Pravdic and Aganovic-Boras also questioned the SCI database for bibliometric purposes and pointed out that it is biased in author studies because the file contains only the name and affiliation of the first author. This could mean that the file omits up to half of the world's important science writers. They examined the bibliographies of dissertations produced at the University of Zagreb and found that 65% of the titles in natural sciences were picked up in SCI, but 94% of them were found in standard databases such as Biological Abstracts and Index Medicus. They concluded that the results of the work "send clear warnings that ISI publications should not be directly used as sources for publication counts of a peripheral country."

Reprint Requests

A curious but interesting approach to understanding the use of language in science was undertaken by Swales (1986). He examined the reprint requests that he had received for an article he had written earlier. While a small data set, he surveyed the requesters and found that all of those that responded to his questions used English as a first or second language.

Book Literature Studies

As noted above, it is important to attempt to lay the trends found in secondary services against any data that can be found to suggest actual production figures of the literature. Data presenting production by language are few and far between. Chapter 6 presents a new set of estimates based on the most heavily cited source for book production statistics, the Unesco *Statistical Yearbook* (see Guha, 1985, for example).

Shearer (1986) attempted to assess the linguistic production of books worldwide. He described the horrendous problems in attempting to assess the book industry output of nations. These problems include varying definitions of a "book" -- from the hundred-page definition from Italy, for example, to Iceland's seventeen pages.

By examining population figures and access to various types of media, Shearer (1985) said that "North Americans lead in opportunities to consume the electronic mass media but the Russians lead in access to the print production of newspapers and the Europeans lead in the production of ... book[s]."

He suggested that the number of new books produced per thousand inhabitants per year was as follows:

Language	No. of new books per 1000 speakers annually
English	.408
Japanese	.370
Portuguese	.152
Arabic	.043
Chinese	.012

He did not offer data for other languages. He proposed that these figures offered an indication of the "productivity" of different languages. Note that these languages are generally *not* those spoken by substantial population groups worldwide.

Starr (1978) also addressed the production of book literature. He stated that the USSR "surpasses all other non-English speaking countries except India in the number of English language titles it produces." This is a significant change in attitude from twenty years ago, when the USSR claimed that the increasing use of English in Asia and Africa was a "sinister plot by the Anglo-Saxon world." If this is true, it is curious why Russian does not appear in the above table.

Regarding Asia and Africa, however, he described a decline in the percentage of English language books published between 1967 and 1972, by two thirds in Asia and a half in Africa.

Justin Bloom (Brady 1986) reported that Japan is increasingly publishing in English, to assert itself as a scientific power, and to build its image. Monthly reports describing the Japanese fifth-generation computer project were published in English. He did not expect the country's years of effort in developing English-language capabilities to be thrown over "just for the purpose of making some point about the internal quality of Japanese research."

In the United States, there may be a resurgence of interest in foreign language titles. Bolle (1987) notes that while the 1960's saw the rapid decline of foreign languages in school curricula, now booksellers and publishers are feeling a demand for them.

Use Studies

A "use study" is an attempt to determine the degree to which materials, particularly in a library, have been used. A popular method is to count books and other materials left on reading tables. Use also can be defined as those works included in a bibliography of a research paper. While these studies do not directly address the question of what is being produced, they do give a sense of the potential demand for materials in non-English languages in specific settings.

The Hutchins study (1971) was a two-year project to determine the extent to which a language barrier exists at the University of Sheffield. Its purpose was to gain guidance on the need for translation services and on the development of the book collection in the library. The project also determined the foreign language competence of the faculty and their literature gathering and citation habits.

The report summarized several studies of world production of scientific and technical periodical article production, and while expressing some skepticism about some of the data, suggested that English content has remained stable or risen somewhat from the late 1950's to the middle sixties overall. The second most important language is Russian.

Studies of journal references show that the primary variables affecting inclusion in a bibliography or list of references are the language of the journal itself, and the subject field. The Louttit study (Brittain, 1970) found that for English psychology journals, English references were the majority (60%), followed by German (15%), French (12%) and other languages (9%). However, Hogeweg-de Haart (Brittain, 1970) found that for the less-commonly spoken Dutch social science journals, only 33.6% of citations were in Dutch, and 51.7% were in English. German contributions were 9.5% and French were at 3.7%.

Hutchins' summation of studies of citation habits was not very helpful, partly because of the variety of approaches taken in each of the studies to determine linguistic content, and the large number of variables needed to understand what was occurring. For example, to make comparisons across fields, one would need to know the language capabilities of the individual authors, the language of the citing article, and, as the author suggests, the full proportion of literature consulted in preparation for the article. Overall, though, references to English works predominated in English language journal articles.

While focusing on science, it is useful to digress for a moment to the humanities for comparison and perspective. The humanities relies more heavily on book titles for research (Cullars, 1988). A set of prize-winning books from *Publishers Weekly* awards was examined to show that foreign language works were cited in these books for 66% of all references, but one or more works on Russian topics skewed the data. The statistics appeared quite high. Later the author asserted that in actuality works cited in languages other than the topic of the book (such as titles to Russian works in a work on Russian literature) account for less than 2% of the total. Foreign language references will be more frequent in the humanities than in the hard sciences because humanities books far more often than scientific ones will cover a country-oriented topic. It is common to find works describing the music or art or literature from France, for example. It is more rare to find a work on French hydraulic research or French chemistry.

Library Holdings

Analyses of collections already built, such as libraries, also reflect use rather than production. However, when the library is a national library, attempting a comprehensive and representative collection, studies of these collections do indicate what the collectors perceive to be representative of the field, or at least the perceived needs of the users.

The British Lending Library

The Wood study in 1966 of the holdings of the British National Lending Library for Science and Technology found holdings in English of 46%, in Russian of 14%, in German of 10%, in French of 9%, in Japanese of 4%, in Spanish and Italian of 3% each, in Polish, Czechoslovakian, and Portuguese of 2% each, and 5% of other languages.

The CONSER Collection

The CONSER (Conservation of Serials) project of the Library of Congress and the Online Computer Library Center (OCLC) represents one of the largest collections of serial titles in the world. Among its over five hundred thousand records, the number of non-English titles has decreased from 25% to just under 20%.

The National Library of Medicine

The National Library of Medicine in 1977 changed its collection policy, placing materials from other countries at a lower priority. This indeed affected the international character of the materials database. This is more fully described in Chapter 5.

Works Missed as an Indicator of Need

The Hutchins study (1971) reported on the degree to which the faculty felt that because of the various deficiencies in access mechanisms provided by the university, and because of the perceived importance of the research in a given language, they were missing significant research. For those languages where there was strong language competence, the percentages were low (but still present). Across disciplines, 9.1% felt they may have missed

French research, and 15.5% felt some German work might have been missed. These countries were considered important by 29.6% and 36.7% of the faculty, respectively. For Russian, 31.5% felt they had missed work of significance, and 25.3% considered the Soviet Union an important country. Only 15.2% considered Japanese important, and 15.9% felt they had missed any work of significance. (In the study, the language and country were equated.)

Translations as a Solution

A solution often proposed for providing access to scientific literatures in other languages is the translation of items in a particular category. These, of course, generate another body of literature which then must be controlled, and access to these materials must be provided. Readers are referred to the vast body of literature regarding the problems of translations and the numbers created each year. Translations here are only mentioned in passing because they are so often considered as a part of the general problem.

Translations are very costly to produce. They can be produced in anticipation of use: for example, cover to cover translations of a specific journal. In this case, however, the translator does not know whether they are going to be used at all.

The high costs of access to translations already completed cause those wishing to use the material to bemoan the lack of coordination to provide access to translations that already exist. Corporate lawyers apparently protect translations done in-house, and a substantial amount of translation is done by the intelligence community. These latter materials are frequently classified, not because of the content of materials or the way in which they were acquired, but because they were translated by an intelligence agency (U.S. House... Availability... 1984). Of course, there is no way to gauge how much literature this represents, or its potential value for industry.

Translations factored into a report by the National Academy of Science (1969) as an "expedient" way to make available to American scientists work published in non-English languages. The report in fact indicated a decline in scientists' language abilities and their desire to achieve proficiency in other languages. One of the chief difficulties cited was that of predicting what should be translated, and for what audience. While custom translations insure that a needed document will get to a user who has a need for or interest in it, there is then the potential for duplication of effort in translation. To avoid this, centers were established such as the Special Libraries Association (SLA) Translations Center in 1953 at the John Crerar Library (later the National Translations Center, now at the Library of Congress) in Chicago. With the delays in production (6-18 months) and then announcement in bibliographic resources, translations clearly are not the best way to stay current with foreign language materials.

Walgren (Brady, 1986) reported that the volume of translations at the Crerar Library in Chicago had in 1980 dropped to an eighth of the effort of that being expended in 1970, due to cuts in primarily National Science Foundation funding.

Another approach is to translate simultaneously with publication, or even before publication. Bernal, in proposing a radical restructuring of scientific communication, addressed the issue of languages. He listed English, French, and German as the three major scientific languages. His vision for scientific communication was the abolition of scientific journals altogether, in favor of massive national-level "pre-print" distribution, through subject centers. Each subject center would also be responsible for selecting and translating appropriate national papers into one or more of the above languages for use in other countries, and for translating externally produced papers into the local language. All abstracts would be published in all three languages.

The third approach is to provide translations at the time of need. This means usually developing a roster of individuals with language skills and the time to produce them, or relying on an outside service which does translations for a business. For the researcher, there is the possibility of delaying the research project, and the addition of further expenses to the budget. And, there is the issue of duplication of effort, when the same document is translated for two different scientists.

One of the most difficult issues is deciding the skills needed for effective translation. The American Mathematical Society, in its translations projects for Russian statistical journals, found that native English speakers provide the best translations (U.S. House... Availability... 1984). This suggested that it was incumbent upon societies to develop the local capacity to deal with foreign languages.

Scholar's Literature Habits

A scientist's citing and publishing practices show how literatures in different languages are being used, and the languages of choice for publishing. As suggested earlier, if a scientist publishes in his own country and in his own language, others may think the work shows an "ignorance of work going on elsewhere." If a scientist publishes outside his country, that is an insult to the country's publishing structure (Inhaber, 1977).

Inhaber's 1977 study of publishing patterns for different countries showed interesting consistencies in insularity for the United States and the Soviet Union. For the first, 80% of a selected sample of United States scientists published in United States sources. For the second, 82% of Soviet scientists published in Soviet sources.

Of particular interest was Herner's study in 1958 of the use of Soviet medical research (Herner, 1958). In this, he was directed to determine the "need for Soviet medical research information among American medical scientists" and, to determine why such information if available was not being used.

The study found that the primary resource used for locating information about Soviet research was through abstracting and indexing services. Further, the study showed that when other approaches to information seeking were examined (for information in general, and for foreign information), when seeking Soviet information, the use of secondary services nearly doubled over the use of "footnotes or other cited references."

The study also demonstrated that there were insufficient means to know about Soviet research in order to use it (which is an interesting response to juxtapose against the emphasis on abstracting and indexing services). Other findings included an inability to "obtain up-to-date Soviet publications," the need for translations, and the need for informal contacts with Soviet scientists.

The recommendations of the scientists participating in the study included an increased coverage of Soviet publications in abstracting and indexing services. This was the strongest and widest recommendation. Less frequently mentioned was the increased dissemination of translated tables of contents, translations, and increased contact between Soviet and American scientists, and increased access to original Soviet publications through some kind of clearinghouse.

As background to the study, the researchers inquired about other languages sought. They found that the languages most often a problem were German (89%) and French (71%), with Spanish (18%) and Italian (15%) falling far behind. The percentages of languages the scientists could read were lower: German and French were reported 87% of the time, Spanish and Italian could be handled 28% of the time each.

Shilling, Tyson and Bernard (1963) studied the use of Russian research by "bioscientists," and suggested that the author's opinion of the general value of Russian research might be a factor in its level of use. This is the same conclusion reached by Herner (1958).

King Research (King, 1976) in an assessment of scientific and technical literature production gathered data on ways by which scientists gather their information. Citations were located in printed indexes in 10.4% of cases, and in computerized literature searches in 0.9% of cases. Given the age of the data and the increasing availability of searches over the last ten years, it is highly probable that this figure is higher today. Of interest also is the fact that 30.5% of items were located by reading other materials. These materials also are, presumably, documented by references, and one might assume that at least 10% of these resources were located in secondary sources, which also are affected by language trends. If there is a language barrier it could be affecting over 11% of the scientist's resources, if not more.

The report also indicates an overall decline in the location of articles via secondary sources from 1973 to 1976. They noted that the fields of the physical sciences and engineering were the largest users of secondary sources, but even these percentages of article identification were less than 20% in these fields. At this time, Lockheed and SDC had forty-six databases available. And, King used cost estimates of $100 per search. While the figures suggest that these services are a less important access route to the literature, the situation has changed dramatically since then.

These data are contradicted by the Hutchins study in 1971. Therein, the methods of keeping up to date in various disciplines were evaluated for the University of Sheffield faculty. The use of abstracts and bibliographies was ranked first by faculty in medicine (87.1%) and dentistry (75%), and second by those in materials technology (78.1%), and pure science (64.7%), for example. The use of such services was lowest in law (16.7%). While the approaches are slightly different, in that keeping up to date is a broader scope of activity than using the materials by citing them directly, secondary services are of more importance to at least British scientists than the King report suggests. It would be stretching the point to suggest that secondary services are providing citations to vast numbers of materials which are just not getting used.

Of interest also is what is done with foreign language materials or references once they are obtained. Of course, this depends heavily upon the researcher's linguistic ability, and the perceived importance of the information contained therein. A large proportion of the faculty are competent in French at Sheffield, so the statistics show that a large proportion would attempt to translate the materials in French themselves. For German, the percentages dropped from the seventies and eighties to the forties and below, and the number of individuals reporting that they would take no further action (i.e., not use the reference at all) rose to a percentage in the teens or higher. Few attempted to translate Russian works, but a greater proportion did attempt to find a translation or summary. The "take no action" group was about half the size of those encountering German materials. For Japanese, there was a greater tendency to take no action at all: the largest "take no action" responses of the four languages were reported for Japanese materials.

Another indicator of the use of different literatures is the appearance of citations in annual reviews of literatures, which are supposed to be a summary of research results and indicator to major literature for a given year in a given discipline. Sengupta (1972) analyzed journal citations in articles in the *Annual Review of Medicine* for the years 1965-1969 inclusively. The author found that English language journal titles were predominant (78.91%), followed by English/French/German multidisciplinary works (12.73%), French language journals (4%), and German language journals (2.18%). Others were less than one percent. This suggests that even if articles in other languages are available, they were not being used or considered important, at least for the preparation of annual reviews in medicine.

For country of publication, Sengupta found the following:

Country	**Percent**
United States	54.55
United Kingdom	17.34
Germany	6.55
Switzerland	5.09
France	4.73
Sweden	2.91
Denmark	2.55
Japan	1.45
Australia	1.09

The remaining four countries comprised less than one percent each of the data set. This parallels Sengupta's findings on languages.

Clark Johnson suggested that publishing patterns in magnetics have changed over the years in Japan. The scientists and engineers there publish first in their own journals, and only secondly in America journals. By implication, they were publishing in U.S. journals before this shift (U. S. House... Availability... 1984). During 1982-84 publications in English have tended to be more "summary articles and overviews," with the primary results being published in Japanese, and not in English at all. Then when the results were published in English (at least for the magnetics journal for IEEE), there was a substantial amount of editing and exchanging of drafts, because the Japanese scientists were not writing in their own language. Dissemination of new results was thus delayed further by one or two years. As Kenneth Furuta suggests, this is an excellent way to keep research findings inside the country and maintain a leadership position.

The Centro de Información Científica y Humanística of the Universidad Nacional Autónoma de México has been concerned about the drain of Latin American scientific writing to the North American journals, and has conducted studies to assess the status of Latin American research writing outside the region's borders (Guinjoán et al., 1982).

An analysis of two years of articles (August 1973-July 1975) by Latin American authors in non-Latin American journals in a wide variety of disciplines showed that:

(a) Brazil contributed 30.5% of the articles published; Brazil and Argentina combined published 57%; and those two plus Mexico, Chile and Venezuela produced 90% of the data set,

(b) Most of the papers were published in United States journals, although in actual first and second place were a Dutch and Swiss journal, respectively, and

(c) One third of the papers were in geophysics, one fifth in chemistry, one fifth in physics and mathematics, nine percent in biology, seven percent in agriculture, and four percent in the humanities and social sciences.

The study did not address the issue of languages, but it did indicate the tremendous pressure to access non-local files. The study did not compare the amount of this literature to that which was published in the region.

Scholar's Language Abilities

It is important to also examine the trends in the ability of scientists themselves to handle materials in languages other than English. And this has been decreasing steadily.

The general lack of or decline in language study, and even knowledge of international affairs by the American population may be in part attributed to the role of the federal, state, and local governments in education and culture, and this is noted widely. In a pluralistic society, these functions are pushed down to local control as far as possible, with the federal government playing as little role as possible. Its role is to serve as a "cultural catalyst," presenting innovative programs as stimuli for local efforts (Keller and Roël, 1978). While there are occasional local programs designed to promote intercultural awareness and connections (sister cities, student exchange programs, and refugee efforts for example), it also generally is not the role of local or state units to maintain an international awareness. Of course there are exceptions, notably in the Southwestern states and in Florida (over immigration concerns) and in Michigan (with the Japanese economic threats to the automobile industry). But unless directly affected, international affairs are left to the federal government. Thus, those in a position to be aware of international dependence among nations have little to do with the educational or cultural system which supports it and develops a country which can participate intelligently and sensitively in the arena.

The study of foreign languages by American students is not as common as it once was, but this trend may be reversing itself. For those college students registered in 1977, French, German, Italian, Russian and Spanish account for 89% of class registrations for languages. Latin and Greek brought the total to 93% (Hayden 1979). Spanish accounted for 46% of total registrations, which Hayden claimed is totally appropriate since the United States is the "fourth largest Spanish speaking country in the world." And, one assumes, it is appropriate for the people of a neighboring country to try to learn the other's language.

Apparently in 1962, 73% of American scientists could handle a second language (Rabkin and Inhaber, 1979). It is doubtful if the same is true today.

While geographic location is a local approach to justify studying a language, it is not a good reason if a culture is developing an educational system for science, as this study will suggest. Spanish is not one of the major languages of science, and not prevalent enough to warrant 46% of all language registrations. In this author's experience at the undergraduate college level, neither the lofty goals of neighborliness, international communication for the country, or preparation for international activities convince students to select a language for study. Rather, it was perceived as the easiest of the available foreign languages to learn.

Gruson (1986) suggests that foreign language study is rising in the United States, but not for scientific reasons. She cites statistics showing more language requirements for admission to college, greater numbers of foreign language courses, and increased attendance in existing classes. Students are learning for business reasons, however, to be able to function in multi-national companies, and not for scientific reasons.

Librarians seem to fare somewhat better. A recent survey of academic librarians (Cain, 1988) reported that 61% of the respondents claimed proficiency in French, German, and/or Spanish. And Conger (1984) certainly assumes that librarians are searching for non-English materials as she describes how to search for (and exclude) them on DIALOG, BRS, and ORBIT. Moynahan (1985) was concerned enough about being able to retrieve Spanish and Portuguese materials on the OCLC Online Union Catalog that she developed a proposal for a stoplist for the system.

After Identification: Problems in Obtaining the Materials

Problems of language follow through in the acquisition of non-English or non-Western works. Once the item has been identified, it must be acquired. Standard practices of reprint requests, interlibrary loan and the like meet with more or less success depending upon the availability of the journal title or book. As more and more libraries face increasing serial title prices, especially from overseas, the tendency is to cease subscriptions to these materials whose immediate use is less easy to document. Not only is it more difficult to identify materials from other countries, it is becoming more difficult to obtain them once they are identified.

Problems for libraries are well documented in the literature. Rosenberg and Whitney (1986), for example, include many examples in their conference papers and proceedings. Gilliam (1987) is quite pointed about the difficulties in working with other countries and identifying their legal materials. She notes that while bibliographical control for France, Germany, and Italy is good, it is less so for Spain, Portugal and Latin America. And the problems extend to corresponding with sources: some Southern Europeans and Latin American sources tend to prefer to correspond in their own languages. Quebec sources apparently will "refuse to recognize" correspondence that is not in French.

She points out that it is still practice to send requests and instructions for shipment and payment in English only, and surmises that this may well be one reason why replies (and materials) are not forthcoming.

In a similar vein, Swales (1988) examined reprint request cards and found that 80% of them were in English. In some cases, German and French also appeared with English. He had only one card in Spanish from Hispanic/Portuguese-speaking areas.

Is English the Language of Research?

The results are inconclusive, according to the studies presented. The chief problem with them is that they tend to assume that the secondary services are representative of the scientific literature that is being produced. As will later be discussed, in some cases this is not a wise assumption. Indeed, in terms of the coverage of many of the services, the evidence contradicts Swales' (June 1985) assertion that while there is a widespread belief that English is the language of research, "it is rare to find any evidence for this belief." It is probably more accurate to say that English is becoming the language of the abstracting and indexing services.

CHAPTER 3

LANGUAGES AND COUNTRIES IN DATABASES IN GENERAL

Early Scientific Communication

The earliest days of formal science in the seventeenth century saw the use of informal methods of communication to discuss ideas. Written exchanges of ideas and discoveries took place through letters, sometimes extensive. Such exchanges were limited because they were generally to friends, not colleagues with whom one might disagree. And they proved to be a slow method for disseminating new ideas. There were simply too many scientists and too much knowledge for letter writing to remain efficient for long (Rose and Rose 1969).

The formation of societies such as the Royal Society of London in 1662 and the Académie Royale des Sciences in 1666 in Paris was stimulated by the concern for improved communication. Members began publishing experimental results and even hired research assistants--one of the earliest examples of funded research on specific problems. Governments often provided funds for the societies. The Académie solicited funds first from the minister Colbert, and then from King Louis XIV, on the grounds that the research was needed "for the reputation of France" (Rose and Rose, 1969). Early on, at least, the French government was sensitive to its scientific standing in the international community. On the whole, though, it was the promise of technological benefits which justified government funding.

One of the leaders in establishing the Académie, Denis de Sallo, founded the *Journal des Sçavans,* the "first scientific journal." Rather than containing original scientific articles (which the later *Philosophical Transactions of the Royal Society* did), the Journal contained news of "new books ... discoveries ... inventions ... machines ... observations ... and legal and ecclesiastical judgements from all countries" (McKie, 1948). Indeed, the journal more closely resembled a contemporary secondary abstracting service than a scientific journal. De Solla Price (1986) asserts that its purpose was "not the publishing of new scientific papers so much as the monitoring and digesting of the learned publications and letters that now were too much for one man to cope with ..."

Science in the sixteenth and seventeenth centuries was a national endeavor, when there was sufficient leisure time to pursue this "non-utilitarian activity," and there were "men intellectually drawn to the objective study of nature" (Storer, 1970). In the middle of the sixteenth century, science began to be institutionalized with the development of associations and research institutions, and it increased its international character (Restivo and Vanderpool, 1974). The goals of science, rather than its origins, gave it its international dimension, Storer suggested. Its growing institutionalization was reflected in its literature growth as well.

In this period, science was united in its use of Latin as a language. Bernal (1939) asserted that around the time of Galileo, nationalism in science was emerging. Some of Galileo's political problems, he suggested, stemmed from the fact that he wrote in Italian rather than in Latin. Further breaks with tradition were made by Stevinus for Dutch, and by Descartes for French. English was slow to come to science: Newton wrote in Latin, rather than in English.

As science developed, different countries served as its focus and center of excellence. Italy served as mentor for France, and France and England played the role for Germany, by providing training, advice and inspiration. The United States and Russia in turn looked to Germany in the twentieth century (Storer, 1970).

Hagstrom (1965) suggested that Germany developed faster than France in science after the turn of the century because science was decentralized and competitive, whereas in France it was centralized and governmentally controlled. Both the United States and England attempted to copy the German model, the former somewhat more successfully.

"The idea that the discoveries of science, whether speculative or useful, were at the disposal of all who could make use of them, and were not to be guarded as private or national secrets, marks the emergence of modern science" (Bernal, 1939).

Publishing in Science

One of the most important reasons for scientific publication is the establishment of the record of what has been accomplished, discovered, or refuted. Publishing in any form provides a written record of what has gone before, so that future research can be built upon it.

The publishing process serves to define the field of knowledge, in that not all papers are accepted for publication, and some may be rejected in the refereeing process. While there are certainly enough journals around that are not refereed, those that are subject to such controls are generally considered to be more prestigious.

Hagstrom (1965) noted that journal articles were the most important form of publication for the development of science (at the time of his writing), whereas books and public addresses were "less serious."

For the individual scientist, the reasons for publishing have been to establish priority over discoveries and new ideas, and to gain recognition for the work done. The latter was publicly not often acknowledged (Hagstrom, 1965). The scientist must also publish in the formal, refereed literature of science. Simple recognition gained through publications written by or about the scientist in the lay press or in the news media was neither sufficient nor appropriate for proper science (Greenberg, 1967).

Evidence of the importance of recognition is paralleled by the practice of citing the work of others within the publication. It is a necessary a norm of science to show how one's research is connected to that of others, and to recognize the contributions of others to one's own work.

Hagstrom (1965) studied ninety scientists and students of science in the formal sciences (mathematics, physics). He suggested that scientific articles were a form of "gift giving" to the scientific community, for the scientist receives no (or little) financial remuneration. Rather, the community bestows prestige and recognition in return. But the prestige is not "publicly acknowledged" as a goal to be sought, or to be expected.

A study of 1,000 scientists in the United Kingdom reaffirmed that refereed journal articles were -- and would remain -- the primary vehicle for scientific information. The author's discussion with members of the U. S. National Academy of Sciences suggested that attitudes were similar in the United States. (Heslop-Harrison, 1983).

Scientists operate under many pressures, one of which is to seek to make known discoveries and new ideas, to "insist upon absolute freedom of communication," and to receive and evaluate critically the work of others (Storer, 1970). There are also the factors of prestige and recognition to spur along publication (Hagstrom, 1965, cited by Restivo and Vanderpool). This is both personal and national, in that governments see themselves in competition for discoveries and advances (Restivo and Vanderpool).

The issue of secrecy in science is an interesting counterpoint to the demands of science to publish. Scientists may well want to protect ideas, for example, before they are released to the general community, so that others will not take them and establish priority before them. Hagstrom (1965) quoted several scientists from his study who felt (at that time) that there was more secrecy generally in Europe than in the United States. By discipline, the respondents listed organic chemistry and biochemistry as the most secretive, physics and molecular biology next, and the formal (physics, math) sciences as relatively open.

Evidence of the pressure to publish is seen in the increase in publishing fraud and hastily assembled results of work. Van de Kamp and Cummings (1987) list 450 articles on "misconduct and fraud in the life sciences" alone. It is not uncommon to see reports of such problems in the daily press.

Electronic Publishing

The idea of publishing scientific work in media other than paper for the scientific community is an interesting and related issue, because of the potential savings for publishers in production costs. It has not proved popular, however, because the community has not worked out a system to include the traditional activities of refereeing.

In the Heslop-Harrison study (1983), scientists were reported to be very much against direct movement toward wholesale dissemination of information in electronic form, replacing print copies of journals for example, if this would cut them off from the Third World scientists.

More recent events such as the increasing number of experiments in electronic transmission of full text materials like those at University Microfilms and those supported by the Apple Library of Tomorrow program suggest that such attitudes may have changed. In Italy, for example, a project was underway in the early 1980's to disseminate doctoral dissertations which were typeset by authors on Apple microcomputers and transmitted electronically to a centralized printing capability for storage on CD-ROMs (Antoni

1985). Increasingly, full text articles are available from electronic database services. Newspaper articles, for example, are available through databases on DIALOG. Full text articles from the *Harvard Business Review* have been available for a number of years. The American Chemical Society offers full articles through its STN International service.

Access to Scientific Literature

Access to literatures is accomplished through a wide continuum of mechanisms, including the informal sharing of documents, to the formal secondary services.

Whether or not access is important depends upon whether there is any evidence of advanced research and development being reported in the literature at all. If all advances are being made in corporate settings and laboratories and reported in proprietary literature, then, access through traditional means will not be possible. A New York University translation project, for example, found that significant advances were taking place in computer science in Japan, in some areas beyond U.S. efforts, but that these were being reported in accessible literature (U.S. House... Availability... 1984). This is contrary to some assertions that Japanese research and development is being done in corporate laboratories and reported only in internal corporate channels.

Indeed, there are factors in the entire document delivery system that can inhibit the acquisition of a document by a scientist. Libraries, postal services, publishers and producing agencies such as research institutes themselves can cause delays in obtaining materials. However the concern of this study is access to the literatures available: finding out about their existence in the first place.

Programs that attempt to obtain the materials themselves may also hinder access. Nwaga (1980) cites the inaccessibility of non-English journals once relevant citations are identified and notes the need for local networks for interlending, funding, and communication systems. Molino (1982) points out that the supporting tools for retrieval, such as thesaurii, are not published in locally-accessible languages. Tell (1984) suggests that the problem may be more political than technical. The lack of educated people in central administrations tends to hinder policy-making that would facilitate the import of information, and, within the library context, a knowledge of languages may be a liability rather than an asset. With strong language skills, a staff member probably would be moved out of the library and into a ministry post. Dickson's (1984) interpretation of the problem area suggests that the literature itself may not be the issue, rather, it is access to the physical technology developed by science and technology, and the human resources that produce that technology, that are actually more attractive to the developing world.

A particular difficulty in gauging access to literatures is the tendency of countries to publish in English, as a second language, to insure that the research results reach the scientific community (U.S. House... Availability...1984). France publishes in French. Japan and the Western European countries know that American (and other) scientists cannot and will not learn the multiplicity of other languages in which research is conducted, and they take it upon themselves to learn English. Therefore, there is a proportion of English-language literature representing research conducted by non-English speaking scientists. This study addressed that issue only tangentially, and focused on the languages in which research results are being published.

The private sector has largely solved access problems to electronics and electrical engineering information in non-English languages, according to Walgren (Brady, 1986). They have simply developed their own channels of access to the needed information. And, they are not likely to share these channels or the information with anyone else. Maryanne Back of the House Committee on Science and Technology, in the same conference, reported that the Committee has noticed that representatives of the private sector are "reluctant" to discuss the channels that they have developed, and this reluctance most likely relates to the proprietary nature of the information sought and obtained.

It is worth noting the recent threats to access to foreign literatures, in particular at the National Technical Information Service (NTIS). NTIS, formerly the Clearinghouse for Federal Scientific and Technical Information, was created as the Publications Board to disseminate German scientific and technical literature captured during and just after the Second World War. Its mission, therefore, historically was to foreign literatures. The Reagan administration (in particular, the Office of Management and Budget deputy director Joseph Wright) was interested in turning NTIS over to a private, commercial organization. He developed his proposal while he was in the Department of Commerce, to which NTIS reports. Arguments from organizations such as the American Library Association have expressed concern over the threat such action poses to NTIS's current breadth of coverage. Concern from other sources was also expressed over the inevitable increase in prices, and the consequent reduced access to both U.S. and foreign materials. In fact, that was a real possibility: both West Germany and Japan have stated that their contributions to NTIS holdings may cease with privatization. They prefer government to government exchanges. Presumably this would also end access through NTIS to Japanese databases for direct searching of Japanese literature.

While the United States federal government appeared to be relatively unconcerned about access to foreign literatures, it has been concerned about foreign access to American scientific literatures. Database companies such as Mead Data Central have essentially been asked to provide agencies such as the Central Intelligence Agency and the National Security Agency with information on their subscribers and their use of the databases offered. They were also asked if they would (not could) "block certain subscribers from using the databases" (Davis, 1987). Chemical Abstracts has been asked to remove "sensitive" information from the databases it offers. In other moves, the government has prevented U.S. scientists from presenting papers on "optics and composite materials" at conferences. The objective was to keep information from "foreign competitors and adversaries." In an act to ward off potential action by the government, Mead removed several federal government databases from its offerings.

The Development of Secondary Services

Problems of access by scientists to foreign literatures have been discussed since at least 1948 at the Royal Society's Scientific Information Conference, in the context of librarians attempting to assist scholars in locating relevant materials and avoiding full translations (Dickson, 1979). There are currently over 3,000 databases. The last 73% of the databases currently available became available in the last six years. This rate of growth is astonishing.

The major online search services do not represent the sole means of accessing these databases: since the early 1970's, countries have been buying and making other arrangements to obtain database tapes for local processing. Aborn et al. (1973), for example, documented the distribution of tapes from the Institute for Scientific Information, and the software necessary to process them.

While online searching was proposed in 1951, it was first demonstrated with the System Development Corporation's (SDC) application of the Protosynthex system on the *Golden Book Encyclopedia* in 1960 (Bourne, 1980). The first national network for online searching was an SDC effort "to provide 13 organizations with access to some 200,000 bibliographic records on foreign technology." In subsequent years, additional efforts were made by the National Aeronautics and Space Administration (NASA), IBM, the National Library of Medicine, the University of Pennsylvania, and Stanford University. Bourne chronicles the development of interactive services for physics, medicine, and foreign technology primarily.

In the 1960's and 70's, government agencies such as the National Library of Medicine, NASA, and the National Institute of Education, and private agencies such as the American Chemical

Society and the American Psychological Association began making their printed indexes to literatures in their discipline available for machine searching. In the 1970's, private agencies in the United States also arose to gather these files, and others, together to make them available from a "single source."

General access to online bibliographic systems began in the early 1970's. OCLC (then the Ohio College Library Center, now the Online Computer Library Center) began service beyond its Ohio service area in 1971. Lockheed's DIALOG system opened its service to general use in 1971-72, building on its government database files. SDC began marketing at about the same time. MEDLARS, the National Library of Medicine's system, began in 1971.

All of this activity became possible at this time because of the widespread availability of computing and telecommunications technologies.

These vendors offer primarily databases developed in the United States. In recent years, however, two services have arrived with a broader geographic base.

The American Chemical Society in 1983 developed cooperative agreements with Fiz Karlsruhe in Germany to jointly develop the Scientific and Technical Information Network (STN). The Japan Information Center of Science and Technology (JICST) joined the effort in 1983. This service currently offers over eighty databases to the scientific and technical market. This is a particularly interesting service, because many of its databases are completely in German, and a familiarity with German is required for their use.

Questel, in Falls Church, Virginia, was originally developed in France, and became available in the United States in 1981. It includes databases on the arts and humanities, but focuses on science and technology and intellectual property. Several of the more than 60 files are multilingual (such as PASCAL) or in French.

One of the tensions in the continuing development of secondary services has been that between the federal government and the private sector, as a political issue and an economic one. Over the past few years, federal policy has been to shift as much information activity as possible to the private sector. Dickson (1984) pointed out that given the desire to control scientific and technical information within the private sector, it would be highly unlikely that a corporation would allow its most valuable information to be included in such a broadly available source. And this is exactly the type of information that is wanted by other U.S. corporations, and by corporations in developing as well as other industrialized countries.

The Role of Secondary Services in Scientific Research

A secondary service is a bibliographic product which attempts to provide its readers with information about literature in its given discipline. It generally includes bibliographic information about the item, and index terms or an abstract or review to describe the contents of the item or to summarize its result. Traditionally published in printed form on a regular basis, secondary services now frequently are available in machine readable form as well.

A secondary service by economic necessity is involved in discrimination of literatures. All items in a body of literature cannot be included: some must be left out. Someone must make a selection of what will be included. Alic for example suggests that this is a fundamental problem in managing access to Japanese literature: providing access to all of it would be prohibitively expensive. He suggests that technical literature is "redundant": that scientists publish and re-publish results in a variety of ways to be sure that they are heard. Thus, total access is not necessary. However, how the selection is made -- and by whom -- are the critical issues. He favors selection by scientists who can assess the value and significance of the work (U.S. House... Availability...1984).

These databases clearly are not the only means of entry into the scientific literature. There are several different ways that one scientist can know about the work of others. He or she can know of it informally, personally, through professional contacts in daily work, in hallway meetings at conferences, and by referral from one contact to another. These contacts are developed in training, by mentors, and also by the conduct of daily work. The scientist knows of the work first hand, or through a colleague. The colleague may discuss his or her work, or share pre-prints of papers he or she has written.

Or, the scientist can use the formal literature and its tools to identify works of relevance. These tools include the primary literature, the published journals, monographs, and reports and other materials which form the literature of a field. These also include the secondary services, such as abstracting and indexing services, which identify the literature of a given type or on a given subject, and compile references into a bibliography. These also include references appended to books and papers.

The requirements for access to these two communication channels are quite different. To access the informal, personal channels, the scientist must be involved in research and work of interest to others. The exchange of information on the personal level is one of barter: the scientist generally gets information to the extent that he is able to share information that is of interest to the giver. And, given the constraints of time, a single individual cannot personally know all the researchers in a given subject area in his own country. It is asking too much to ask a scientist to know of research in all the countries in the world.

To have access to the printed literature, the scientist needs access to the primary sources. Some of these will be contained in personal collections and those of colleagues. Others will be contained in libraries. As the informal communication process cannot support access to all individuals, the formal, printed process cannot support access to all primary materials. Hence, the emergence of secondary services, which function as guides to the total body of literature available. In printed form, these generally are available in libraries. In electronic form, they are available to anyone with a microcomputer or terminal, a modem, and functional telephone service (and the money to pay for the information), anywhere in the world.

It is this characteristic of availability, anywhere in the world, that makes electronic secondary services of particular interest, and which lend importance to the study of the nature of these databases.

In recent years, more and more countries other than the United States have been developing national databases for monographic materials and specifically in areas of science and technology. The proportion of non-English materials in these databases naturally is far greater than in U.S.-based files. The critical issue is access to these files. Many European databases are available through Questel, a vendor similar to the U.S. services of Lockheed Information Services, which offers databases through its DIALOG system, and Bibliographic Retrieval Services and its BRS system. These services offer hundreds of bibliographic files on virtually any subject, with a complex searching language for retrieving information. Only in the last two years has Questel been readily searchable through EasyNet, a U.S.-based gateway service. EasyNet is a "user-friendly" service in which the user simply enters a subject as a query, and does not have to know the specific database to be searched.

The Import of Whole Databases, System Linkages

Another access route to non-English language materials is a deliberate acquisition of foreign files, or access to systems through telecommunications links. The National Technical Information Service (NTIS), for example, has made available files on science and technology, medicine, government reports, and new technology and products from the Japan Information Center of Science and Technology.

Databases in the United States also sell tapes overseas for local processing. The Institute for Scientific Information (ISI) has been doing so for some years (Aborn, 1973).

In 1985, Chase Econometrics added Germany Database, and NewsNet added Japan Computer Industry Scan, on research and development and marketing in the industry.

It has been reported that the first Spanish (that is, one from Spain) publicly available database became accessible in 1984. Earlier ones, covering drugs for example, were "restricted to certain groups of users." The database BARATZ covers topics in economics, politics, and labor (First Spanish... 1984).

Garvey (1979) broadly delineated two types of scientists: the major and the minor. The first operates at the forefront of a field and is making the discoveries which advance the field. The minor scientist is verifying (or refuting) the discoveries of the major. While both are necessary for the advancement of a field, their information needs, and channels, are radically different. The major scientists are in informal contact with others in relatively the same position. They exchange ideas informally, personally, and have relatively little use for published information as a critical source, when working in their own area of expertise. The minor scientists, on the other hand, do not have such extensive connections, and must rely on printed resources. The secondary services serve these scientists primarily, or the major scientists when they are delving into areas with which they are not familiar.

A question that necessarily arises is the degree to which the secondary services actually represent the total literature (qualitatively and quantitatively) of a discipline. Engineering Index conducted a survey in 1982, and estimated that of the 9,000 Japanese scientific and technical periodicals, there were about 4,000 in engineering. About 600 or 14% of them were covered by secondary services. Two-thirds of these 600 titles were in English. Engineering Index at that time covered 130 journals, yielding about 8,500 titles per year. Landau readily admitted that these journal titles were the easiest to acquire. Research staff examined the potential coverage in areas of interest would encompass 1000 titles. They determined the costs for a comprehensive service, and for a service covering a subset of the whole. These costs (for the comprehensive service, over a half a million dollars a year) were prohibitive. Market research yielded insufficient interest in a subscription to launch such a project (U.S. House... Availability... 1984).

A study by Robert Gibson of General Motors assessed the Japanese literature relevant to his company and found that about 20% of the total titles published were indexed in Western sources. While this might be presumed to question the findings of this study (in that the secondary services in fact *do not* adequately represent the actual literatures in these languages) the preferred interpretation is that even though there may well be more out there, there is no access to them through secondary services, and for all intents and purposes they are invisible.

Languages at the Database Level

While the analysis of the eight databases provides detailed information on the types of materials published by different countries in different languages, as background and context it will be useful to present a gross analysis of the proportion of English as represented by databases available worldwide.

To that end, two resources have been examined. First, the 1982 edition of Martha Williams' *Computer-Readable Databases: A Directory and Data Sourcebook* presented a listing of as many "publicly available," "word-oriented" databases as the editors could identify. Language of records and country of origin were two of the elements which were available for analysis. Second, this database was expanded and has been kept up to date as the Database of Databases on DIALOG, which will serve as the second resource.

Both of these sources include information about the languages of the documents contained in the databases they describe, about the records in those database, and about the countries of origin of the databases.

The Printed *Directory of Databases*

The *Directory* lists 773 separate databases. The method of compilation of the *Directory* affected the completeness of data, which was not unusual or unexpected. According to its introduction, the directory was compiled from database producer responses to a survey, with follow up by telephone and letter by the editors. Therefore inevitably the data were incomplete in some instances: all producers did not provide all of the information requested. Further, no significance could be assumed from the order of languages when provided, because it appeared that no significance was requested: and some of the entries list a large number of languages in alphabetical order. Some producers simply reported that they included "multiple" languages, or materials in "over 50" languages from "over 150 countries" or the like. Despite these influences, the directory nonetheless represents the state of the knowledge about databases at the time of compilation, December 1981.

The producers did not standardize the names of languages or countries in the entries, nor was there correlation between the languages given and the names of countries represented.

Languages of Documents in Databases

Entries in the *Directory* include the languages of the documents contained in the databases. This gives a broad picture of the languages claimed by a large number of databases. Table 3-1 presents the languages cited and the frequency in which they were cited.

Table 3-1, Languages of Documents in Databases, 1982

Language	Frequency of Mention	Percent of Databases Citing
English	470	65.3
Multiple	243	33.8
French	194	27.0
German	128	17.8
Russian	53	7.4
Japanese	38	5.3
Italian	36	5.0
Spanish	36	5.0
Swedish	20	2.7
Dutch	17	2.3
Norwegian	13	1.8
Hungarian	11	1.5
Polish	11	1.5
Portuguese	11	1.5
Danish	10	1.4
Finnish	10	1.4
Czech	6	.8
Scandinavian langs.	6	.8
Afrikaans	5	.7
Chinese	4	.5
Romanian	4	.5
Arabic	3	.4
Bulgarian	3	.4
Persian	3	.4
Turkish	3	.4
Serbo Croatian	2	.3
Basque	2	.3
Catalan	2	.3
Greek	2	.3
Slavic languages	2	.3
Slovak	2	.3
African	1	.1
Albanian	1	.1

Table 3-1, Continued

Language	Frequency of Mention	Percent of Databases Citing
Amerindian	1	.1
Anglo-Saxon	1	.1
Breton	1	.1
Croatian	1	.1
Estonian	1	.1
European	1	.1
Flemish	1	.1
Hebrew	1	.1
Icelandic	1	.1
Indonesian	1	.1
Irish	1	.1
Latin	1	.1
Latvian	1	.1
Lithuanian	1	.1
Malay	1	.1
Nordic	1	.1
Pacific	1	.1
Provencal	1	.1
Romanish	1	.1
Slovenian	1	.1
Vietnamese	1	.1
Welsh	1	.1
Yugoslavian	1	.1

Total number of databases: 718

Total number of languages mentioned: 59

The table shows the wide variety of languages that are included, at least on occasion. While there were a substantial number of languages mentioned, no claims for comprehensive coverage of a language's materials, in general or in the specific subject area of the database, were made. Two databases, Library and Information Science Abstracts (LISA) and the Library of Congress MARC (Books-All) database, skewed the results by reporting twenty-three and forty one languages, respectively. The Chemical Abstracts Service Source Index database did not report individual languages, its entry stated that it included materials in "over 50 languages." And, a third of the databases simply reported "multiple languages," alone or along with the other languages they did specify. Therefore it is a somewhat inaccurate presentation of the actual availability of materials in different languages: however, it is quite accurate in presenting "known languages covered" in the *Directory.* In other words, if a researcher wanted to know what databases included materials in Arabic, without specifically asking each and every database producer, he or she could turn to this directory and find that there are three.

Nonetheless, it is readily apparent that English language materials were included by the largest number of databases. It is highly probable that those reporting "multiple languages" also include English among others, which would have raised the percentage to 99% of the databases including some English materials. Other than English, French, German, and Russian, the remainder of languages are given negligible attention.

Minimum English Content

Given the information in each entry, it was possible to compute an estimate of the amount of English in some of the databases represented.

For 433 databases a "minimum amount" of English could be determined. Of these, 163 (38%) were totally in English. Forty (9%) could be described with certainty as containing no English at all. The remaining 230 contained varying proportions of English.

Table 3-2 shows the increasing proportion of English in these databases. Those databases that were 50% or more in English represented 287 databases, or 67% of the total.

Table 3-2: Minimum English Content in Databases

Percentage in English	Number of Databases	Cumulative English Total
0%	40	
1-9%	10	10
10-19%	14	24
20-29%	20	44
30-39%	19	63
40-49%	43	106
50-59%	39	145
60-69%	18	163
70-79%	40	203
80-89%	16	219
90-99%	11	230
100%	163	393

Databases with No English

By examining entries, it was possible to determine those which did not (because of direct statements) or very probably did not (by subject or country of origin) contain any English records. This very conservative review identified 95 databases. This includes the earlier 40 databases which directly stated that they contained no English.

Table 3-3 represents these databases, by their language and country of origin. There were only 4 instances of databases produced entirely in languages not local to the country of origin: the Public Affairs Information Service Foreign Language Index, produced in the U.S.; the U.S. Patent Information Service from Japan; the philosophy database Philosophie 2 and the Program Information System from West Germany.

Table 3-3: Languages of Databases and Country of Origin

Database Language	Country of Origin	Number Produced
European	Luxemburg	1
Finnish	Finland	1
French	France	11
	Canada	9
	US	1
German	Germany	13
	US	1
Italian	Italy	10
	US	1
Japanese	Japan	1
Multiple	Germany	8
	Austria	1
	Finland	1
	France	1
	Netherlands	1
	Sweden	1
	Switzerland	1
Norwegian	Norway	2
Portuguese	US	1
Spanish	Chile	1
	US	1
Swedish	Sweden	3

Country of Origin of the Databases

Generally, one expects the language of a database to be the language of its country of origin. For those databases which contained no English, this proved to be true. Italian databases were in Italian; French databases were in French.

Those databases entirely in English were from England, the U.S., Canada and Australia. One was from West Germany.

Table 3-4 shows the distribution of the country of origin for databases with some English, by percentage of English contained. Note that generally, as the percentage of English increases, the number of countries whose national language was not English declined sharply at the 60% level.

Table 3-4: Percentage of English and Country of Origin

1-9% English:
Produced in:
Australia, Belgium, Chile, France, Norway, Sweden, United States, West Germany

10-19% English:
Produced in:
Canada, Finland, France, Norway, Sweden, West Germany

20-29% English:
Produced in:
Belgium, Canada, England, France, Sweden, United States

30-39% in English:
Produced in:
Canada, Denmark, France, Italy, Norway, Switzerland, United States, West Germany

40-49% in English:
Produced in:
Austria, England, France, Netherlands, Norway, Scotland, Sweden, United States, West Germany

50-59% in English:
Produced in:
Australia, Canada, England, France, Japan, Sweden, Switzerland, United States, West Germany

Table 3-4, Continued

60-69% in English:
Produced in:
Canada, England, Sweden, United States, West Germany

70-79 % in English:
Produced in:
England, Netherlands, Sweden, United States, West Germany

80-89% in English: Produced in:
England, South Africa, United States

90-99% in English:
Produced in:
Australia, Canada, Finland, United States

Country of Origin of Source Materials

Directory entries also included the countries of publication of the source materials included in each database. These gave a sense of the potential for non-English language materials, and the degree to which it could be suspected that English language materials were produced by those for whom English is not their first language. Table 3-5 presents the countries, and the frequency in which they were cited.

Table 3-5: Countries of Source Documents, 1982

Country of Source Materials	Number of Times Cited	Percent of Databases Reporting
Worldwide	280	38.9
US	319	44.0
Canada	115	16.0

Table 3-5, Continued

Country of Source Materials	Number of Times Cited	Percent of Databases Reporting
England, Gr. Britain, UK	100	13.9
Germany (all references)	83	11.6
France	64	8.9
UK	59	8.2
Japan	41	5.9
Germany (sic)	40	5.6
West Germany (sic)	35	4.9
England	31	4.3
Australia	29	4.0
Russia	23	3.2
European	22	3.0
Sweden	22	3.0
Italy	19	2.6
Norway	18	2.5
Netherlands	15	2.0
Denmark	14	1.9
Switzerland	14	1.9
Austria	11	1.5
Finland	11	1.5
Belgium	10	1.4
Great Britain	10	1.4
Hungary	10	1.4
Poland	10	1.4
India	9	1.2
New Zealand	9	1.2
Spain	9	1.2
Czechoslovakia	8	1.0
East Germany (sic)	8	1.0
Israel	7	.9

Table 3-5, Continued

Country of Source Materials	Number of Times Cited	Percent of Data-bases Reporting
Rumania	7	.9
Yugoslavia	7	.9
Egypt	5	.7
Argentina	4	.5
Brazil	4	.5
Bulgaria	4	.5
Holland	4	.5
Hong Kong	4	.5
Portugal	4	.5
Scandinavian countries	4	.5
Cuba	3	.4
Iran	3	.4
Ireland	3	.4
Mexico	3	.4
Philippines	3	.4
Saudi Arabia	3	.4
Singapore	3	.4
Zambia	3	.4
Chile	2	.3
Greece	2	.3
Iceland	2	.3
Kenya	2	.3
Malaysia	2	.3
South Africa	2	.3
Taiwan	2	.3
Turkey	2	.3
Venezuela	2	.3
Western Europe	2	.3
Zimbabwe	2	.3

Cited once: Africa, Antarctica, Asia, Azerbaidzhan, Bahrain, Bangladesh, Bolivia, Caribbean countries, China, Circumpolar countries,

Colombia, Costa Rica, Cyprus, Greenland, Guyana, Indonesia, Iraq, Jamaica, Jordan, Korea, Kuwait, Latin America, Libya, Luxemburg, Malawi, Malta, Monaco, Mongolia, Nigeria, Pakistan, Qatar, Rhodesia, Thailand, Uganda, United Arab Emirates. These were each cited by one percent of all databases, or just once.

These data also showed the dominance of attention to United States materials and those of Canada, the United Kingdom, Germany and France. Japan was the only non-European country that was mentioned by more than 5% of the databases. Russia was mentioned by only 3.2% of the databases, which was surprisingly low. The data were distorted somewhat by the large number of databases reporting "worldwide" coverage, and databases such as the Chemical Abstracts Service Source Index which reported covering materials from "over 150 countries."

Languages and Country Sources of Documents Compared

The USSR was disproportionately mentioned in comparison to its placement in the language table: twenty-three databases report covering its materials, yet fifty-three databases report covering materials that are *in* Russian. Note also that forty-one databases reported covering Japanese literature, and a surprising thirty-eight databases reported covering materials in Japanese. This suggested that some databases may have been limiting coverage to English language materials from Japan.

The most frequently mentioned languages and countries were ranked, in descending order, as follows:

Language	**Country**
English	US
French	Canada
German	UK
Russian	Germany
Japanese	France
Italian	Japan

Spanish	Australia
Swedish	Russia

Overwhelmingly, the literatures of the predominately English-speaking countries were contributing to the high ranking of English as a language. France and Germany, and German and French are in reverse order: in part this could be attributed to a Canadian contribution to the French language materials. Japan and Russia and their languages are also in reverse order. Japan was cited as a source of materials almost twice as many times as Russia, yet Russian as a language was cited half again as often as Japanese.

The 1985 Database of Databases

The DIALOG file called the Database of Databases contained largely the same type of information as the *Directory*. This file was searched to yield information comparable with that above. This database contains 2,473 records, 1,700 records more than the printed Compendium. The online database had nearly triple the number of records of databases available.

Country of Origin of Databases

Entries for the databases included information about their country of origin. These countries, and their frequencies, are presented in Table 3-6 below.

Table 3-6: Databases by Country of Origin

Country	Number of Records	Percentage of Total
USA	1,589	64.2
Canada	163	6.6
France	152	6.1
United Kingdom	81	3.2
Japan	79	3.2

Table 3-6, Continued

Country	Number of Records	Percentage of Total
England	72	2.9
Germany	70	2.8
Australia	51	2.1
Italy	40	1.6
Spain	33	1.3
Sweden	28	1.1
Netherlands	22	.08
Denmark	14	.05
Belgium	13	.05
Norway	10	.04
Switzerland	10	.04
Luxemburg	9	.03
Austria	6	.02
China	6	.02
Finland	6	.02
Great Britain	3	.01
South Africa	3	.01
Chile	2	.008
Thailand	2	.008
Bermuda	1	.004
Indonesia	1	.004
Ireland	1	.004
Israel	1	.004
Malaysia	1	.004
Poland	1	.004
Singapore	1	.004
Wales	1	.004
Yugoslavia	1	.004
Total	2,473	

The majority of databases were produced in the United States. Conspicuous in its absence was the Soviet Union.

Languages of Records in Databases

The database contained information on the language of records in the database (as opposed to the language of the documents to which the records refer). Titles may be translated, transliterated or otherwise presented differently from the source documents.

Table 3-7 following lists these languages and their frequencies. The total does not yield the total number of the records in the database because a single database could use more than one language for its records.

Table 3-7: Principal Languages in the Database of Databases

Language	Number of Records	Percent of Total Records
English	1,770	76.6
French	180	7.8
Japanese	70	3.9
German	64	2.8
Spanish	43	1.8
Kanji	39	1.6
Italian	37	1.6
Kana	23	.09
Swedish	18	.07
Dutch	16	.06
Danish	12	.05
Scandinavian	7	.03
Norwegian	6	.02
Chinese	5	.02
Multilingual	5	.02
Finnish	3	.01
Portuguese	2	.008
Latin	1	.004
Polish	1	.004

Table 3-7, Continued

Language	Number of Records	Percent of Total Records
Romaji	1	.004
Slovene	1	.004
Total	2,305	

English dominates the entire series of records. The equivalence of Kanji with Italian was curious. Again, Russian was conspicuous by its absence.

Language of Source Materials in Databases

The database also contained information about the languages of the documents represented by the records. These represent a wider variety of languages. Table 3-8, using the data in the database itself, presents these figures:

Table 3-8: Languages in Databases, 1986

Language	Number of Records	Percent of Data-bases Reporting
English	468	65.1
Multiple	243	33.8
French	194	27.0
Russian	64	8.9
Italian	60	8.3
Spanish	57	7.9
Japanese	41	5.7
Dutch	28	3.8
Swedish	27	3.7
Danish	18	2.5
Portuguese	17	2.3
Norwegian	15	2.0
Finnish	12	1.7

Table 3-8, Continued

Language	Number of Records	Percent of Databases Reporting
Polish	12	1.7
Chinese	11	1.5
Hungarian	11	1.5
Czech	10	1.3
German	10	1.3
Romanian	5	.7
Scandinavian	5	.7
Slavic	5	.7
Afrikaans	4	.5
Bulgarian	4	.5
Arabic	3	.4
Croat	3	.4
Farsi	3	.4
Serbo	3	.4
Hebrew	2	.3
Kana	2	.3
Kanji	2	.3
Slovak	2	.3
Turkish	2	.3
Belgian	1	.1
Catalan	1	.1
Flemish	1	.1
Gaelic	1	.1
Greek	1	.1
Indonesian	1	.1
Serbocroatian	1	.1
Slavonic	1	.1
Slovene	1	.1
Welsh	1	.1
Yugoslavian	1	.1

Total number of languages: 43

Changes Between 1982 and 1986

Languages in Databases

Two comparisons were made regarding languages. One compared the languages of the documents in the 1982 *Directory* with those of the *Database of Databases*. Table 3-9 presents this information.

Table 3-9: Languages Mentioned, 1982 and 1986

Language	1982	Percent Citing	1986	Percent Citing
English	470	65.0	468	65.1
Multiple	243	33.8	243	33.8
French	194	27.0	194	27.0
German	128	17.8	10	1.3
Russian	53	7.4	64	8.9
Japanese	38	5.3	41	5.7
Italian	36	5.0	60	8.3
Spanish	36	5.0	57	7.9
Swedish	20	2.7	27	3.7
Dutch	17	2.3	28	3.8
Norwegian	13	1.8	15	2.9
Hungarian	11	1.5	11	1.5
Polish	11	1.5	12	1.7
Portuguese	11	1.5	17	2.3
Danish	10	1.4	18	2.5
Finnish	10	1.4	12	1.7
Czech	6	.8	10	1.3
Scandinavian langs.	6	.8	5	.7
Afrikaans	5	.7	4	.5
Chinese	4	.5	11	1.5
Romanian	4	.5	5	.7
Arabic	3	.4	3	.4
Bulgarian	3	.4	4	.5

Table 3-9, Continued

Language	1982	Percent Citing	1986	Percent Citing
Persian	3	.4	2	.3
Turkish	3	.4	2	.2
Basque	2	.3	0	0
Catalan	2	.3	0	0
Greek	2	.3	1	.1
Serbo Croat.	2	.3	1	.1
Slavic langs.	2	.3	5	.7
Slovak	2	.3	2	.3
Kana	0	0	2	.3
Kanji	0	0	2	.3
African	1	.1	0	0
Albanian	1	.1	0	0
Amerindian	1	.1	0	0
Anglo-Saxon	1	.1	0	0
Breton	1	.1	0	0
Croatian	1	.1	11	1.5
Estonian	1	.1	0	0
European (sic)	1	.1	0	0
Flemish	1	.1	1	.1
Hebrew	1	.1	2	.4
Icelandic	1	.1	0	0
Indonesian	1	.1	1	.1
Irish	1	.1	0	0
Latin	1	.1	0	0
Latvian	1	.1	0	0
Lithuanian	1	.1	0	0
Malay	1	.1	0	0
Nordic	1	.1	0	0
Pacific	1	.1	0	0
Provencal	1	.1	0	0
Romanish	1	.1	0	0
Slovak	1	.1	2	.3
Slovenian	1	.1	1	.1

Table 3-9, Continued

Language	1982	Percent Citing	1986	Percent Citing
Vietnamese	1	.1	0	0
Welsh	1	.1	1	.1
Yugoslavian	1	.1	1	.1
Belgian	0	0	1	.1
Catalan	0	0	1	.1
Farsi	0	0	1	.1
Gaelic	0	0	1	.1
Romanian	0	0	5	.7
Serbo	0	0	3	.4
Slavonic	0	0	1	.1

This table suggested that there have been minimal changes in the languages represented over the last four years: except in the case of German, which has dropped from 128 to ten. It did return to 160 in 1989.

An examination of the file in 1989 found few changes. All of the languages were represented in the same frequency as found in the 1986 search of the database, with a few exceptions. Yugoslavian increased from 1 to 2 mentions, and German did so from 10 to 160.

A second comparison was made between the languages of the records in the two sources. This information represents the primary language(s) of the databases themselves. Table 3-10 presents these data on the following page.

Table 3-10: Languages of Records in Databases, 1982 and 1985

Language	1982	% of Records	1985	% of Records
English	470	34.2	1770	76.6
Multiple	248	17.8		
French	194	14.0	180	7.8
German	128	9.3	64	2.8
Russian	53	3.8		
Japanese	38	2.7	70	3.9
Italian	36	1.9	37	1.6
Spanish	36	1.9	43	1.8
Swedish	20	1.4	18	.07
Dutch	17	1.2	1	.06
Norwegian	13	.09	6	.02
Hungarian	11	.08		
Polish	11	.08	1	.004
Portuguese	11	.08	2	.008
Danish	10	.07	12	.05
Finnish	10	.07	3	.01
Czech	6	.04		
Scandinavian langs.	6	04	7	.05
Afrikaans	5	.03		
Chinese	4	.03	5	.02
Romanian	4	.03		
Arabic	3	.02		
Bulgarian	3	.02		
Persian	3	.02		
Turkish	3	.02		
Serbo Croatian	2	.01		
Basque	2	.01		
Catalan	2	.01		
Greek	2	.01		
Slavic languages	2	.01		
Slovak	2	.01		
African	1	.007		
Albanian	1	.007		

Table 3-10, Continued

Language	1982	% of Records	1985	% of Records
Amerindian	1	.007		
Anglo-Saxon	1	.007		
Breton	1	.007		
Croatian	1	.007		
Estonian	1	.007		
European	1	.007		
Flemish	1	.007		
Hebrew	1	.007		
Icelandic	1	.007		
Indonesian	1	.007		
Irish	1	.007		
Latin	1	.007	1	.004
Latvian	1	.007		
Lithuanian	1	.007		
Malay	1	.007		
Nordic	1	.007		
Pacific	1	.007		
Provencal	1	.007		
Romanish	1	.007		
Slovak	1	.007		
Slovenian	1	.007		
Vietnamese	1	.007		
Welsh	1	.007		
Yugoslavian	1	.007		
Kanji			39	1.6
Kana			23	1.6
Romaji			1	.004
Romanji			1	.004
Slovene			1	,004

The table shows a disturbing absence of Russian and Hungarian, and a significant decrease in the variety of languages available.

Country of Source Documents in Databases

The *Database of Databases* does not include information about the country of the source documents, so this important feature of databases could not be calculated or compared.

Country of Origin

The next comparison was made for the country of origin of the databases. The table following, Table 3-11, outlines these differences. It presents the countries cited in the 1982 *Compendium,* and those reported in the 1985 Database, to show the number of databases produced by these countries.

Table 3-11: Country of Origin of Databases, 1982 and 1985

Country	Number Cited	% of Total	Number Cited	% of Total
US	379	52.7	1589	64.2
England	81	11.2	72	2.9
United Kingdom			81	3.2
Canada	68	9.4	163	6.6
France	44	6.1	152	6.1
W. Germany	38	5.3	70[1]	2.8
Japan	14	2.0	79	3.2
Italy	12	1.7	40	1.6
Netherlands	10	1.4	22	.08
Sweden	10	1.4	28	1.1
Norway	7	1.0	10	.04
Australia	6	.8	51	2.1
Finland	4	.5	6	.02
Austria	3	.4	6	.02
Luxemburg	3	.4	9	.03
Belgium	2	.2	13	.05

[1]1985 data do not separate the Germanies: this is simply "Germany" and represents both.

Table 3-11, Continued

Country	Number Cited	% of Total	Number Cited	% of Total
Chile	2	.2	2	.008
Denmark	2	.2	14	.05
Scotland	2	.2	0	0
South Africa	2	.2	3	.01
Switzerland	2	.2	10	.04
India	1	.1	0	0
Spain			33	1.3
China			6	.02
Thailand			2	.008
Bermuda			1	.004
Indonesia			1	.004
Ireland			1	.004
Israel			1	.004
Malaysia			1	.004
Poland			1	.004
Singapore			1	.004
Wales			1	.004
Yugoslavia			1	.004

Note the increasing dominance of the United States as a producer and the continued dominance of the English speaking countries. Between 1982 and 1985, France maintained its position, other countries simply fell in their presence. The number of countries producing databases at all (even if only one) increased dramatically.

Summary

The growth of databases over the last thirty years has been dramatic. An examination of growth and change patterns in the 1980's, however, suggested that while databases were produced by an increasing number of countries, their language patterns have hardly changed at all, even though three times the number of databases were identified between 1982 and 1985. English remained the dominant language at the database level, and while there was a wide variety of languages represented, there was no evidence of intense or comprehensive coverage for any of them. Individual languages were scattered thinly among the databases identified.

CHAPTER 4

LANGUAGES AND COUNTRIES IN EIGHT MAJOR DATABASES

Preliminary Notes

This chapter presents the language and country data found in the eight major databases selected for the study. The data were collected primarily by searching the databases on the DIALOG system. (For details, see Appendix 5.) For Oceanic Abstracts the records were downloaded and processed on the Michigan Terminal System at the University of Michigan. As it turned out, this was fortunate. This is the only file in the set that indicates the languages of summaries of documents that can be compared with the primary language in the record. Those data have since been removed from the file. While somewhat older than the updated information, these are the only data on summary languages analyzed for the project.

To reiterate the selection criteria, each of the databases was chosen because it intends to provide international coverage of its respective discipline, it contains records over a sufficient time span (ideally back to 1970) to provide a historical perspective, and it contains "tagged" fields such as language and date of publication that can be searched directly.

BIOSIS Previews

The BIOSIS file, produced by BioSciences Information Services of Philadelphia, Pennsylvania, is the major source for information on biology, life sciences, and associated subdisciplines.

Types of Documents

The file contains just five basic document types:

Journal Article	2,998,273
Conference Paper	1,031,972
Conference Proceedings	36,300
Book	29,517
Patent	7,076

Languages Represented

BIOSIS lives up to its international character. Its records represent 54 named languages:

Table 4-1: BIOSIS Languages

Language	Number of Records
English	3,498,994
Russian	188,888
German	93,142
French	82,634
Japanese	79,885
Spanish	34,789
Italian	27,357
Chinese	15,737
Portuguese	13,689
Polish	11,622
Bulgarian	7,815
Korean	6,108
Romanian	5,035
Czech	4,534
Norwegian	4,290
Serbo-Croatian	3,960
Danish	3,844
Ukrainian	3,179
Hungarian	2,782

Table 4-1, Continued

Language	Number of Records
Netherlandish (sic)	2,341
Belorussian	1,987
Slovak	1,934
Swedish	1,516
Turkish	1,320
Hebrew	1,216
Azerbaijani	814
Slovenian	724
Afrikaans	700
Finnish	670
Greek	352
Indonesian	233
Georgian	189
Serbian	164
Thai	149
Arabic	97
Icelandic	72
Catalan	67
Farsi	53
Latin	51
Hindi	45
Macedonian	34
Armenian	28
Estonian	28
Swahili	21
Albanian	17
Latvian	16
Cambodian	3
Croatian	3
Esperanto	3
Malay	3
Lithuanian	1
Provencal	1
Tadzhik	1
Tibetan	1
Total	4,103,138

This table is presented to show the variety of languages represented in the BIOSIS database. The original data were gathered in 1986, but this chart has been updated to reflect the database in June of 1989. During the interim, most languages represented increased their numbers by approximately 30-40%. Of the more substantially represented languages, Chinese and Korean showed the highest increases overall, nearly doubling their frequency. Most of the languages at the lower end of the scale showed much lower increases. There were three additional documents in Latin. To provide a more manageable study, just the twelve languages most frequently cited were examined, which represent 4,060,660 records out of a total of 4,103,138, or 99% of the records with an identified language. English itself represented just over 85% of those records studied, so the remaining evaluated languages represented approximately 14% of the set. The languages other than the twelve studied totalled less than 1%. The languages studied were English, Russian, German, French, Japanese, Spanish, Italian, Portuguese, Chinese, Polish, Bulgarian, and Korean.

Language Trends

BIOSIS is a difficult file to assess, because the records were not consistently coded with the language of the documents until the late 1970's. Therefore, the later years have been described. All of the languages appeared to have increased since coding began in 1976, simply because coding itself increased, not because there were more materials published in a given language for a given year.

In this section for BIOSIS and for the other databases, charts for selected languages are included to illustrate a point. Charts for all languages are included in the Appendix.

For each year of publication, more and more records were present in the database, as seen in the chart at the top of the next page.

BIOSIS Records:
All Records, by Year of Publication

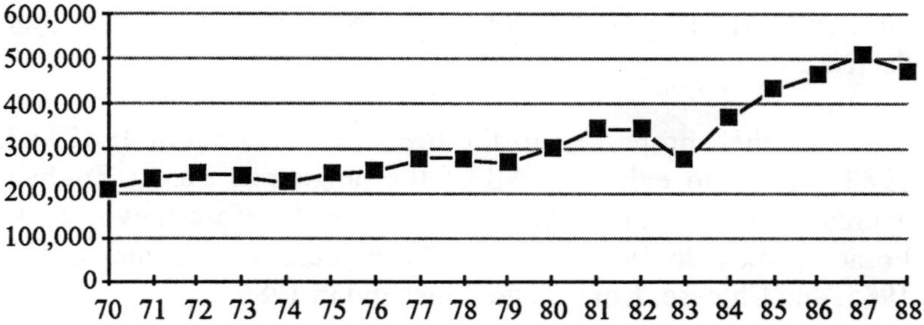

In terms of their numbers, most of the languages (Bulgarian, French, German, Italian, Japanese, Korean, and Polish) reflected a similar pattern: a steady increase in numbers until 1982-83, a steeper or less steep decline for those years, and recovery and increase thereafter. That for French serves as an example:

BIOSIS Records for French Documents:
Records by Year

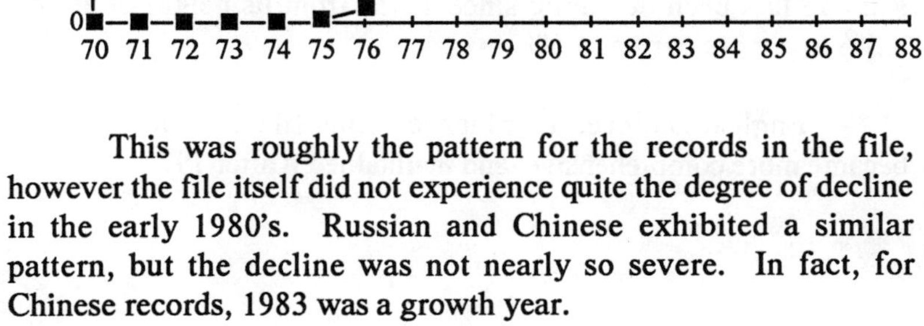

This was roughly the pattern for the records in the file, however the file itself did not experience quite the degree of decline in the early 1980's. Russian and Chinese exhibited a similar pattern, but the decline was not nearly so severe. In fact, for Chinese records, 1983 was a growth year.

Portuguese and Spanish lost ground in the 1980's, unlike the rest, and declined in 1986-87.

English mirrored the pattern of the database itself, except that the number of records for documents published in 1982 did not decrease.

In their proportion of the file, the decreases for 1982 and 1983 were also evident. All of the languages except English decreased their representation, but peaked before they did so. French peaked in 1984, German, Portuguese and Spanish did in 1985, and Chinese, Korean and Polish did in 1986.

Using French again as an example:

BIOSIS Records for French Documents:
Percent of All Records, by Year

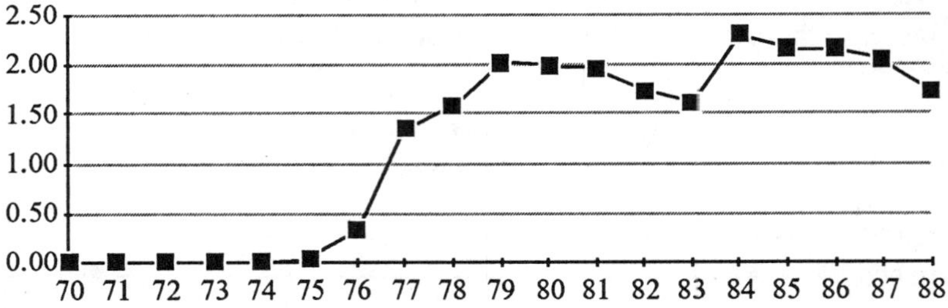

Japanese was the only language with sufficient strength in 1988 to suggest that it would develop a strong upward path. Russian had been declining since 1979, from its height of around 5.5% to a more current 3.5%.

English achieved dominance of the file as coding activity became more comprehensive, and maintained it after 1977.

Chemical Abstracts

The CA Search file, including Chemical Abstracts and associated files, is produced by the Chemical Abstracts Service in Columbus, Ohio. It is the pre-eminent file for chemical research information. It is complemented by several other files on DIALOG and on its own system, STN (The Scientific and Technical Information Network), available from Chemical Abstracts.

Types of Documents

Chemical Abstracts records were found to be of the following specified types:

Conference Proceedings	65,267
Technical report	33,354
Dissertation	22,704
Book	14,446

The remainder of the records are journal articles. Patents and chemical structures are additional sources of important information for chemical research, and these are found in different databases.

Languages Represented

The number of languages in the file demonstrates its international character:

Table 4-2: Chemical Abstracts Languages

Language	Number of Records
English	4,858,579
Russian	1,382,889
Japanese	682,562
German	423,524
French	204,816

Table 4-2, Continued

Language	Number of Records
Chinese	84,255
Polish	80,244
Italian	51,513
Spanish	46,791
Czech	45,464
Romanian	31,917
Bulgarian	27,491
Hungarian	26,732
Portuguese	17,038
Ukrainian	16,133
Korean	15,736
Slovak	14,082
Netherlandish (sic)	10,241
Serbo-Croatian	8,280
Swedish	5,111
Croatian	4,092
Turkish	2,674
Belorussian	2,422
Finnish	2,276
Danish	1,966
Norwegian	1,552
Slovenian	1,547
Serbian	1,148
Greek	981
Vietnamese	744
Georgian	673
Indonesian	675
Hebrew	668
Armenian	611
Albanian	589
Afrikaans	576
Lithuanian	511
Azerbaijani	420
Hindi	305
Latvian	294
Persian	277
Macedonian	259
Estonian	246
Thai	156
Arabic	119
Malay	79

Table 4-2, Continued

Language	**Number of Records**
Basque	63
Esperanto	59
Cambodian	51
Catalan	51
Interlingua	26
Icelandic	24
Turkmen	14
Swahili	9
Gujarati	7
Faeroese	5
Mongolian	5
Panjabi	5
Gaelic (Scottish)	4
Moldavian	4
Sindhi	3
Tamil	3
Burmese	2
Irish	2
Kazakh	2
Tadzhik	2
Welsh	2
Latin	1
Total	8,059,602

This table shows the wide-ranging variety of languages present in the database (68, including English). The table was constructed originally in 1986, and updated with data gathered in June of 1989. In the intervening period, only Chinese changed its relative position, moving from seventh position to sixth. Most of the more frequent languages, with 10,000 or more documents, increased their numbers by around 5-10%. Chinese was one exception, with an increase of 27%. Japanese was the other, with an increase of 15% overall. For the lesser languages, with numbers less than approximately 2,000 records, the statistics were virtually stagnant. Hindi and Arabic both fell into this latter category, two languages that are spoken by large populations.

At this count, the languages represented 8,059,602 records and were more linguistically diverse than most databases in that only 60% were in English. In order to examine 99% of the file, records for 15 languages needed to be studied: English, Russian, Japanese, German, French, Polish, Chinese, Italian, Spanish, Czech, Romanian, Bulgarian, Hungarian, Portuguese, and Ukrainian.

Language Trends

Chemical Abstracts was unusual in the set of eight databases in that four of the languages examined showed very erratic patterns that could not be classified as increasing or decreasing. These were Hungarian, Polish, Portuguese and Romanian. They did not represent large numbers of records, however.

Five of the languages showed patterns of increase: English, Chinese, Czech, German, and Japanese. Japanese was interesting in that the number of records nearly tripled between 1978 and 1979, and then continued to generally increase:

Chemical Abstracts Records for Japanese Documents:
Records by Year

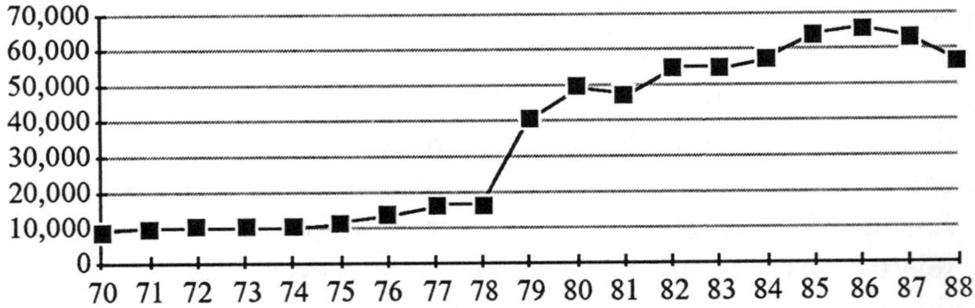

Spanish records increased up until 1986, but then started to decrease in frequency. While this could have been caused by delays in adding records to the database, that explanation is unlikely because Spanish is generally easier to manage than, say, Chinese, Czech or Slovak, which increased between 1986 and 1987.

Records for documents in Bulgarian, French, Italian, Russian, and Ukrainian showed a pattern of decrease. The Russian pattern was not as certain because records increased in the early seventies, and a stable number of records were added for materials published until the middle 1980's:

Chemical Abstracts Records for Russian Documents:
Records by Year

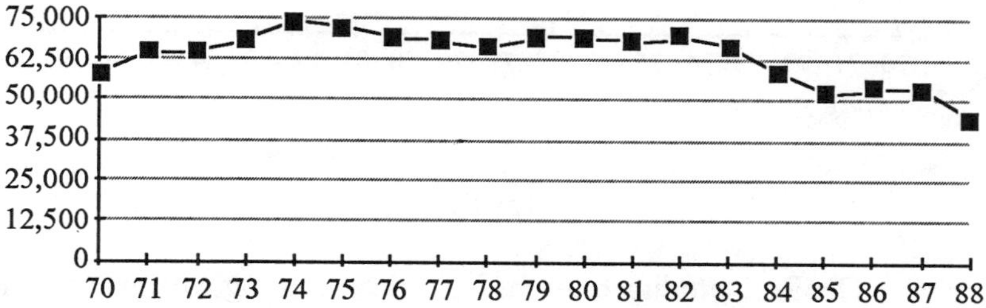

The French pattern shows the same decline in the same years.

As a proportion of the file, four languages were just too erratic to classify, or they were stable and showed no decisive pattern: German, Portuguese, Romanian, and Spanish. One could make a case for increases for German and Spanish, but not easily.

More languages were declining as a percentage of the file than their raw numbers would have suggested. The list included French, Bulgarian, Hungarian, Italian, Polish, Russian, and Ukrainian. The largest group in the set was Russian, whose records decreased by half. The chart for Russian is shown at the top of the next page.

Fewer languages showed the increases that their raw numbers might suggest: just English, Chinese, Czech, and Japanese. Czech records did not make up one percent of the database. It was interesting to note the Japanese increase from around three percent to almost 14% of the file.

Chemical Abstracts Records for Russian Documents:
Percent of All Records, by Year

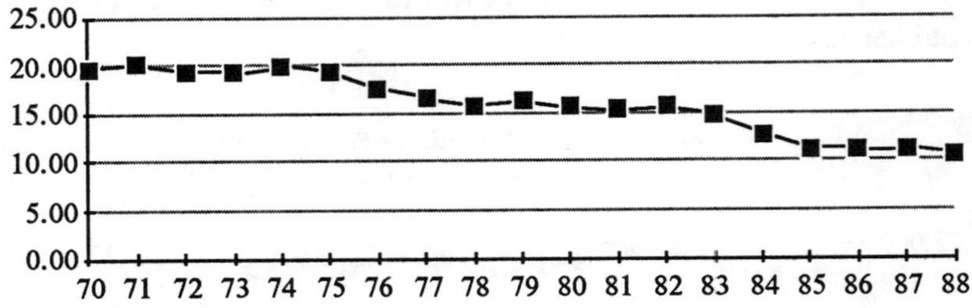

GeoRef

GeoRef, produced by the American Geological Institute in Alexandria, Virginia, contains references to worldwide literature on geology, geography, mineralogy, petrology, and associated subjects.

Types of Documents

The GeoRef file contains literature of the following types:

Serial	753,056
Conference Publication	196,397
Book	107,261
Thesis	32,607
Report	28,604
Map	28,066

Languages Represented

The file contains a diverse set of 53 identified languages, as seen in the table beginning on the next page.

Table 4-3: GeoRef Languages

Language	Number of Records
English	784,993
Russian	121,401
French	55,893
German	45,826
Spanish	21,365
Chinese	16,326
Japanese	15,598
Polish	8,168
Italian	8,076
Portuguese	5,013
Czech	4,806
Romanian	3,355
Ukrainian	3,054
Hungarian	2,469
Bulgarian	2,229
Slovakian	1,563
Croatian	1,319
Korean	933
Swedish	922
Serbian	712
Finnish	670
Dutch	660
Turkish	649
Norwegian	473
Greek	377
Slovenian	265
Polyglot	220
Danish	211
Lithuanian	182
Albanian	162
Catalan	157
Hebrew	136
Afrikaans	125
Georgian	113
Arabic	106
Flemish	106
Estonian	105
Indonesian	89
Icelandic	72
Azerbaijani	49

Table 4-3, Continued

Language	Number of Records
Macedonian	29
Eskimo	28
Malaysian	27
Thai	26
Persian	9
Hindi	4
Welsh	4
Armenian	4
Latin	2
Basque	1
Belorussian	1
Latvian	1
Vietnamese	1
Total	1,110,093

The number of English-language records also may be over-represented because of record ideosyncrasies.[1]

Records for those languages accounting for 10,000 records or more were examined. These were English, Russian, French, German, Spanish, Japanese and Chinese. These represented 95-99% of the file, depending upon how the records are counted.

Like the other files, the first snapshot of this database was taken in 1986. Since that time, the numbers of several languages have increased dramatically. Chinese again was the most productive, with an increase of 53%, followed by Japanese with 23%. German, Spanish, Portuguese, Czech, and Slovak increased between 17-20% each. Russian, the largest non-English language, increased by just 12%. A number of less frequent languages had substantial increases (such as Korean at 30% and Catalan at nearly 50%) but they are represented by fewer than 1,000 records.

[1]The number of English records probably also includes records with English summaries, although the text may be in a different language.

Language Trends

GeoRef had been increasing the number of records for each publication year until 1984, when fewer and fewer records were added each year:

GeoRef Records:
All Records, by Publication Year

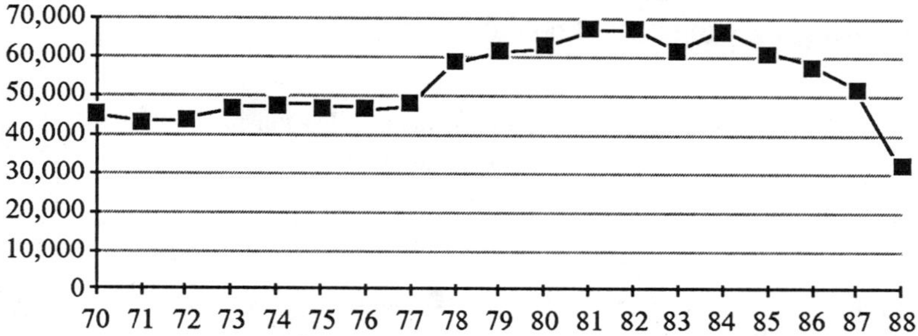

French, German, and Spanish show a strikingly similar pattern, in that the number of records for documents published in the 1970's was roughly the same each year. There were substantial increases around 1981, and then the number of records for each declined, both in raw numbers and as a proportion of the file. A chart of the French records serves as an example:

GeoRef Records for French Documents:
Records by Year

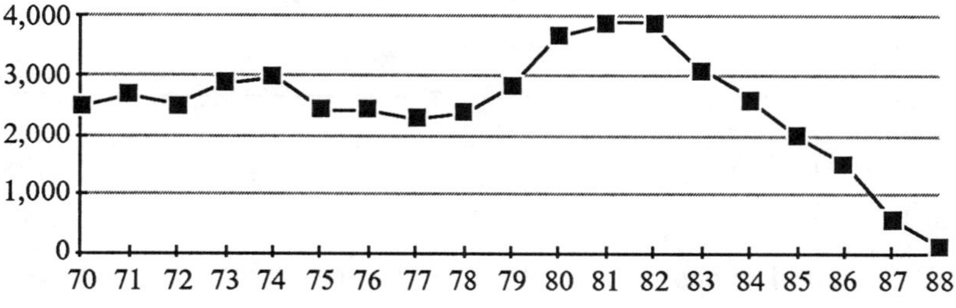

Russian showed a more steady decline, without the increase in the early 1980's.

Japanese and Chinese, on the other hand, increased their strength, particularly in the middle 1980's. While they are not a large percentage of the file (in 1987, just under 2% for Japanese and 4.5% for Chinese), their numbers nearly doubled between the two decades. While all of the 1988 records have probably not been added to database, the proportion of Japanese for that year increases, showing a good potential for growth.

GeoRef Records for Japanese Documents:
Records by Year

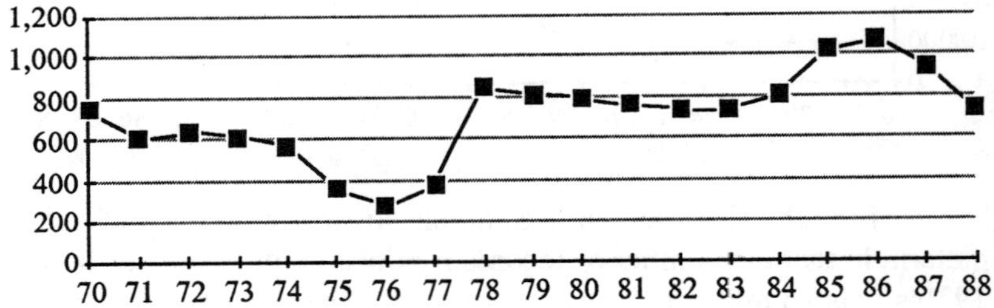

GeoRef Records for Chinese Documents:
Records by Year

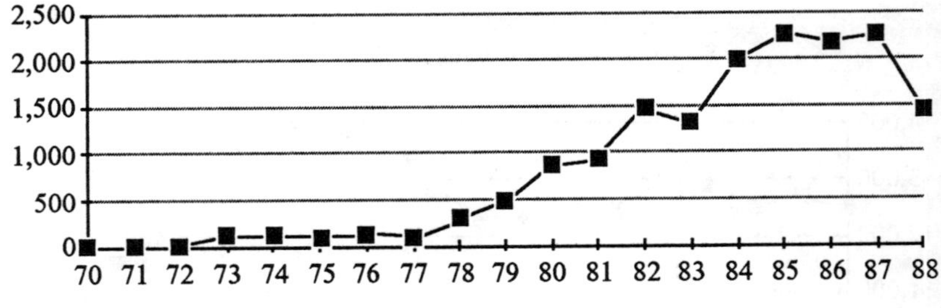

The file has become more and more dominated by English records, even though their actual numbers showed a pattern of increase until 1984 and then decline. The chart of the English proportion of the file shows that even these lower numbers are an increasing proportion.

GeoRef Records for English Documents:
Percent of All Records, by Year

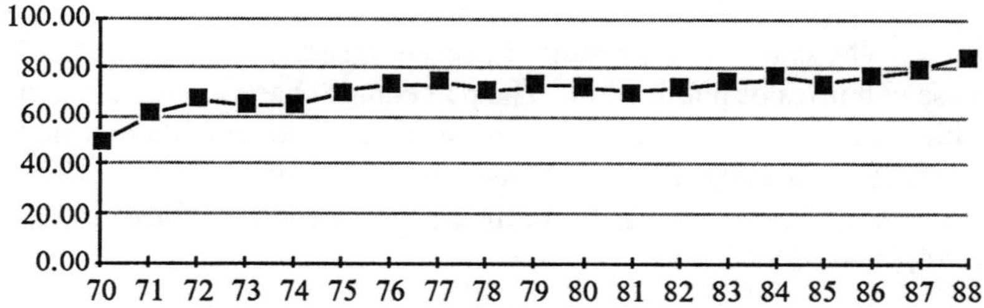

Since the database generally increased the number of records for each publication year, it could be expected that the proportions of records in different languages would also increase. While this happened for English and Chinese, it did not for the other languages.

The charts did not show a significant pattern for the proportion representation of the different languages (such as a common strong year). English and Chinese were clearly on the increase. Spanish and Japanese, while in a declining pattern, fluctuated around a single percentage point. French fell from around 5-6% to around 3%. While the single year may be an aberration, at the very least the proportion is fluctuating around a stable average. German is dropping from around 5% in the early seventies to around 4% in the middle seventies. Russian is clearly on the decline, from roughly 15% in the early seventies to around 8% or 9% in the early eighties.

The patterns of the charts for raw numbers and percentages for each language are virtually mirror images of each other. The only exceptions of note were the charts for English, and for its 1983/84 years. While the raw numbers of documents in English were decreasing, their percentage of the database was still increasing, continuing the trend for the 15-year period.

Countries of Publication Represented

It was possible to identify some of the countries of publication from the GeoRef database.

The charts in Appendix 3 present trends in the numbers of these countries of publication. The percentage charts are to be taken with some caution, because the proportion of documents that had identified this information increased over the years. For the 1970 documents, 3% had it. In 1976, the percentage rose to 66%, and in 1984, it was 88%.

The file contained records from 151 countries:

Table 4-4: GeoRef Countries

Country	Number of Records
United States	330,173
Union of Soviet Socialist Republics	75,069
International	55,133
United Kingdom	42,108
Canada	34,781
Germany, Federal Republic of	30,152
France	29,738
Netherlands	25,091
Japan	19,552
China	18,958
India	14,047
Australia	12,845
Czechoslovakia	8,614
Italy	7,464
Spain	7,335
Poland	7,256
"Varies"	6,605
German Democratic Republic	5,899
Austria	4,369
New Zealand	4,083
Brazil	3,996

Table 4-4, Continued

Country	Number of Records
Argentina	3,909
South Africa	3,455
Switzerland	3,378
Ukrainian SSR	3,188
Hungary	2,953
Sweden	2,945
Bulgaria	2,464
Romania	2,428
Mexico	2,376
Yugoslavia	2,217
Norway	2,137
Belgium	2,055
Denmark	1,905
Israel	1,451
Finland	1,429
Greece	1,341
Venezuela	1,188
Portugal	1,042
Turkey	982
Korea, Republic of	861
Malaysia	837
Egypt	786
Taiwan	776
Chile	757
Thailand	721
Colombia	720
Cuba	592
Pakistan	443
Ireland	401
Indonesia	389
Saudi Arabia	373
Morocco	372
Peru	335
Costa Rica	293
Hong Kong	286
Nigeria	275
Iraq	274
Tunisia	260
Philippines	247
Papua New Guinea	246
Iceland	244

Table 4-4, Continued

Country	Number of Records
Jamaica	220
Algeria	165
Zimbabwe	165
Bolivia	144
Monaco	136
Botswana	127
Southern Rhodesia	120
Cyprus	115
Ecuador	111
Albania	102
Fiji	99
Senegal	92
Iran	89
Luxemburg	87
Syria	65
Byelorussian SSR	58
Libya	58
Singapore	56
Zambia	54
Jordan	53
Kuwait	53
Ethiopia	49
El Salvador	47
Dominican Republic	41
Kenya	39
Nepal	39
Guatemala	37
Uruguay	37
Malawi	36
Nicaragua	32
Upper Volta	32
Mozambique	29
Bahamas	28
Bangladesh	28
Tanzania, United Republic of	27
Cameroon	25
Haiti	24
Ivory Coast	20
Liechtenstein	20
Madagascar	20
Puerto Rico	19
Tonga	19

Table 4-4, Continued

Country	Number of Records
Oman	18
Zaire	17
Ghana	16
British Solomon Islands	15
Solomon Islands	15
Vietnam, Republic of	15
Guam	14
Namibia	14
Sri Lanka	14
Bermuda	12
Sudan	12
Central African Republic	10
Guyana	10
Niger	10
Surinam	10
Yemen	10
Angola	9
Greenland	9
Swaziland	8
Burundi	6
Hebrides	6
Honduras	6
New Hebrides	6
Liberia	5
Trinidad & Tobago	5
Panama	5
Burma	4
Afghanistan	3
New Caledonia	3
Paraguay	3
Rwanda	3
Barbados	2
Guinea-Bissau	2
Korea, Democratic People's Republic	2
Qatar	2
Somalia	2
United Arab Emirates	2
Antigua	1
Chad	1
French Afars & Issas	1
French Polynesia	1

Table 4-4, Continued

Country	Number of Records
Gabon	1
Khmer Republic	1
Lebanon	1
Lesotho	1
Sierra Leone	1
Uganda	1
Total	799,728

The list represents 799,728 records, about half of the 1.4 million records in the database. While 151 countries are in the table, nearly half (70) contributed fewer than a hundred records each. Ninety percent of the file was contributed by nineteen countries, and ten accounted for 75%.

The statistics even more dramatically demonstrated the international growth of the file over the last few years. The original data from 1985 included all of these countries but three, and they produced fewer than twenty-five records all together in the final count. However, the rates of growth of records from the countries contributing more than 1,000 records have been substantial. China leads the list with a growth of 64%, followed by the Ukrainian SSR with 54%. Six countries (Federal Republic of Germany, the Netherlands, Austria, Australia, New Zealand, and Brazil) had growth between 30-40%. Fourteen countries grew between 10-20%: the United States, United Kingdom, Canada, Japan, India, Czechoslovakia, Poland, Spain, the German Democratic Republic, South Africa, Sweden, Romania, Brazil and Portugal, as well as the records designated "International."

Country of Publication Trends

The charts fell into several patterns, which could be generally described as decreasing in numbers. Those from the Soviet Union and those coded "International" peaked in numbers in the late 1970's.

GeoRef Records for Documents from the USSR:
Records by Year

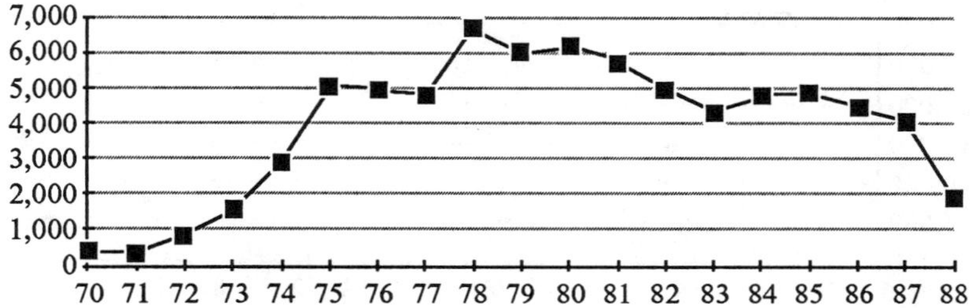

Those from France and the Federal Republic of Germany peaked in 1981-82, and those from the Netherlands in 1983-84.

GeoRef Records for Documents from France:
Records by Year

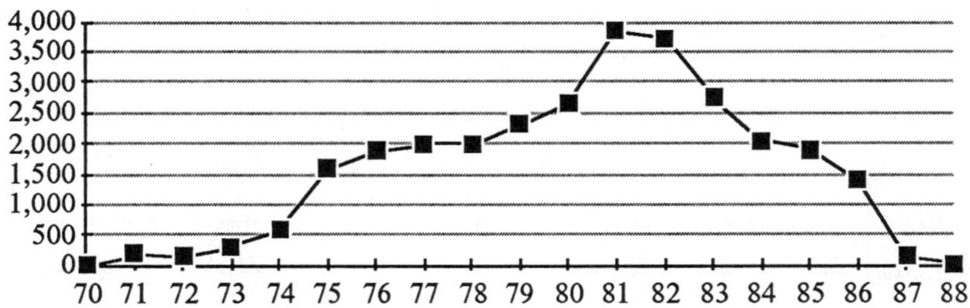

Those from the United States, United Kingdom, and Canada very roughly follow the pattern of the file itself, showing increasing numbers in the late 1970's, stability, and then decline in the later 1980's. Japan also roughly falls into this category. The chart for the United States is shown at the top of the next page.

GeoRef Records for Documents from the United States:
Records by Year

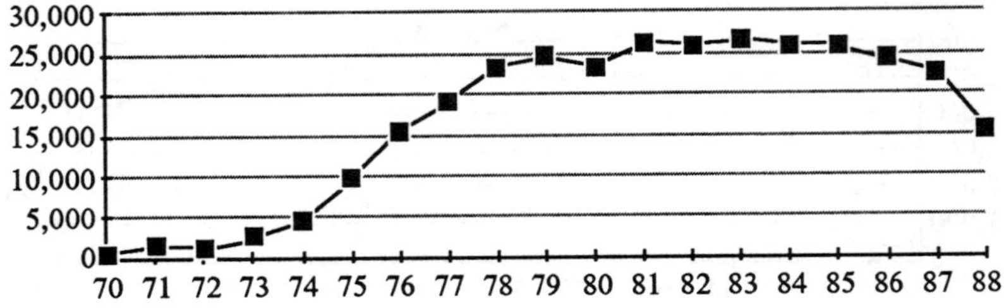

As a percent of all records, those four countries (and those identified as "International") increased their proportion of the file. Even in 1988, with incomplete data, the United States proportion for the first time reached 50%, and the Japanese proportion (although a small 3-3.5%) also increased.

GeoRef Records for Documents from the United States:
Percent of All Records, by Year

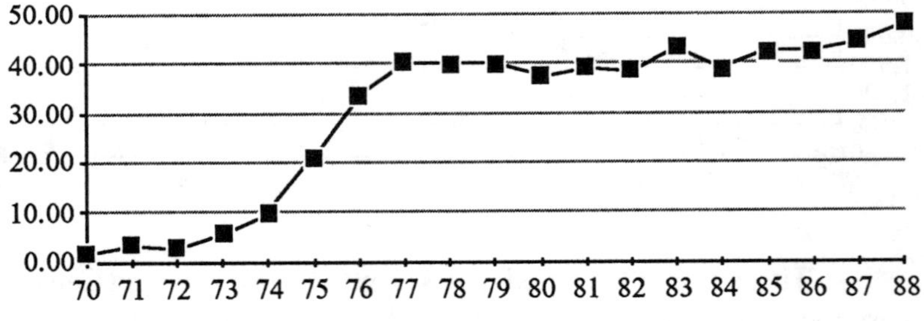

This is not unexpected, as United States materials would be the first to arrive at the offices of the database provider and would be more rapidly processed and added to the file. Similar remarks can be made about Canadian materials. It should take no more effort to receive materials from France than the United Kingdom, however the former are less and less in evidence, and the latter more and more.

As in other files, the number of documents from China showed a pattern different from the rest: a reasonably steady growth throughout the period, in numbers and in proportion of the file. The 1988 figure can probably be explained by the difficulty in handling the materials and adding them to the file.

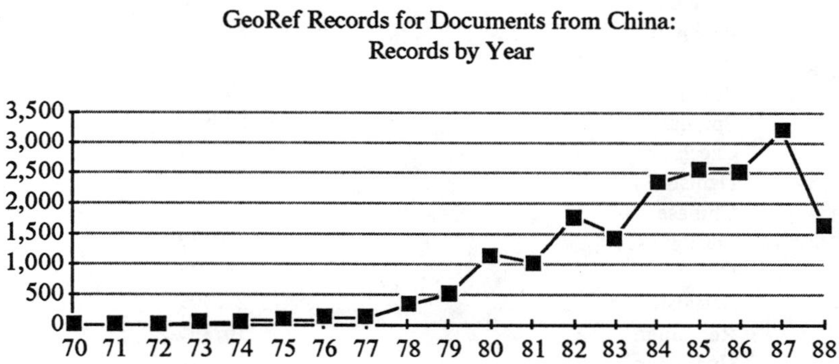

MEDLINE

The MEDLINE database represents the world's major file for information on research and clinical medicine. It is produced by the National Library of Medicine, in Bethesda, Maryland.

Types of Documents

The file contains number of miscellaneous categories of materials, such as reviews and special lists. It identifies 21,659 monographs, the remainder of the file is composed of journal articles. Other materials important to the field of medicine, such as books, are found in separate files.

Languages Represented

The 39 languages represented in the file are diverse, as seen in the table beginning on the top of the next page.

Table 4-5: MEDLINE Languages

Language	Number of Records
English	4,098,295
German	416,752
Russian	380,632
French	291,176
Japanese	192,786
Italian	133,164
Polish	90,727
Spanish	83,887
Czech	44,812
Danish	24,105
Chinese	23,793
Portuguese	23,229
Dutch	22,098
Swedish	21,409
Romanian	16,995
Hungarian	15,474
Norwegian	15,022
Serbo-Croat. (Roman)	12,938
Bulgarian	12,183
Slovak	11,214
Ukrainian	9,307
Hebrew	7,750
Serbo-Croat. (Cyril.)	7,451
Finnish	7,345
Turkish	1,976
Greek, Modern	1,970
Afrikaans	1,831
Korean	1,686
"Multilingual"	1,392
Thai	485
Slovene	387
Icelandic	142
Arabic	127
Persian	112
Armenian	38
Catalan	11
Indonesian	5
Georgian	1
Latin	1
Total	5,972,708

The initial data for the file were gathered in 1986, and updated in June of 1989. In that period, most languages increased between 8 and 16 percent. Of special note was Serbian, which increased by 33%, and Chinese, which increased by 43%. For some reason, the number of records in Czech decreased by 20%.

For this study, the top fifteen languages were examined: English, Chinese, Czech, Danish, Dutch, French, German, Italian, Japanese, Polish, Portuguese, Romanian, Russian, Spanish, and Swedish. These constituted 98% of the database. MEDLINE is more linguistically diverse than other files, it takes a substantial number of languages to make up the bulk of the database. However, it includes a lesser number of total languages (thirty-nine) than, for example, Chemical Abstracts.

Language Trends

Each year, MEDLINE contained an increasing number of records for each year of publication in the database:

MEDLINE Records:
All Records, by Year of Publication

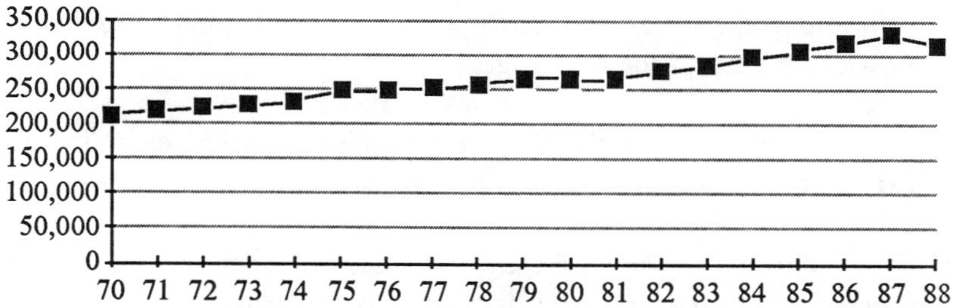

In actual numbers, the languages represented in the database showed a predictable three different patterns: increasing, decreasing, or simply stable.

Those decreasing included French, Romanian, German, Polish, Czech, and Swedish. The Swedish records, however, increased in 1988, contrary to its pattern from 1978. The decreasing trends for the other languages are pronounced. The chart for German will serve as an example:

MEDLINE Records for German Documents:
Records by Year

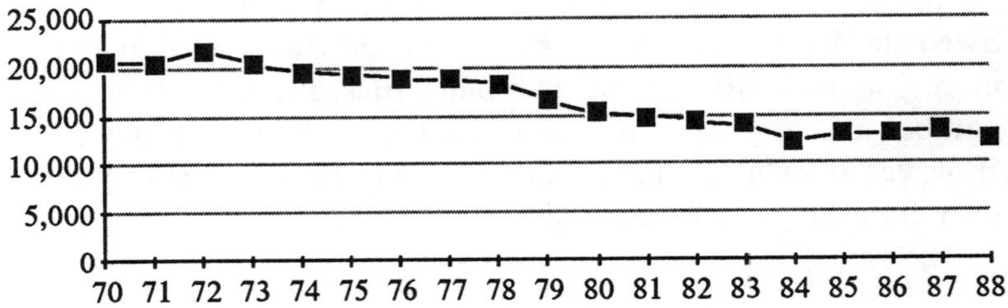

English, not unexpectedly, increased its numbers dramatically, from a contribution of around 125,000 a year in 1970 to 250,000 in 1987. Other languages showing increasing trends included Chinese, Danish, Dutch, Japanese, and Spanish. Chinese was most interesting in that its contribution was negligible until 1978, but in it 1979 began an upward movement such that by 1985 the number of records present for each year had tripled. Its chart is shown at the top of the next page.

Portuguese and Russian were stable across the years. Italian was just too erratic to classify with confidence.

The proportion of the total database that the languages represent showed a similar classification of patterns.

MEDLINE Records for Chinese Documents:
Records by Year

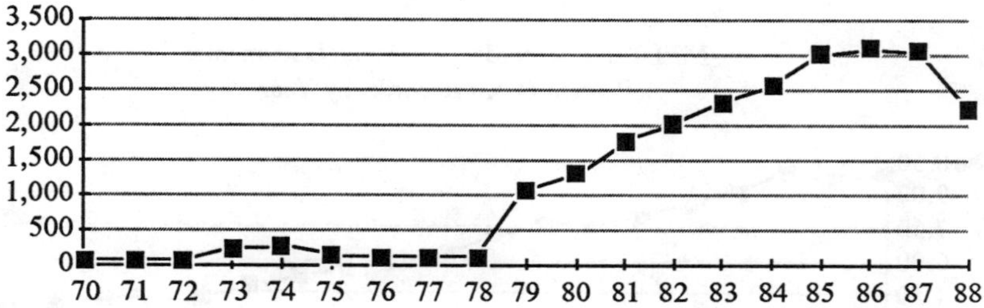

The only languages showing increases in their proportion in the file were Chinese, Japanese, and English. The determination for Japanese is not made with much confidence, because in 1970 it represented 4% of the file, and in 1987, 3%. Between 1970 and 1980, it was 3% or below, and between 1980 and 87, it was above 3%. These small numbers (Chinese was always less than 1%) did not indicate an increasing proportion of non-English materials overall. This was further substantiated by the English pattern of continued increase in the file:

MEDLINE Records for English Documents:
Percent of All Records, by Year

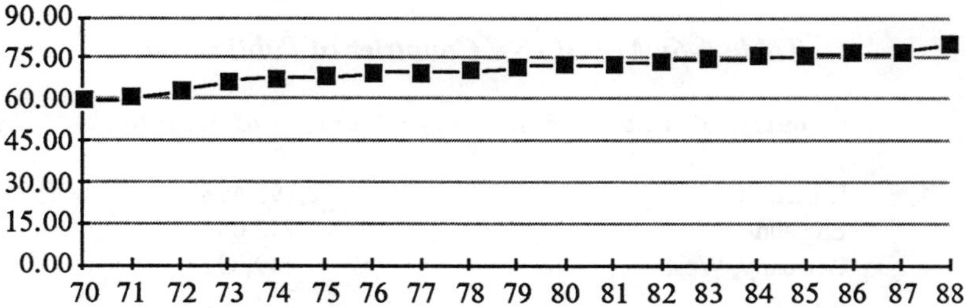

Those languages decreasing in proportion in the file included Czech, French, German, Italian, Polish, Portuguese, Romanian, Russian, and Swedish. The change for French, German, and

Russian combined was dramatic: in 1970, they constituted 27% of the file, but in 1987, they made up 12%. German serves as the example.

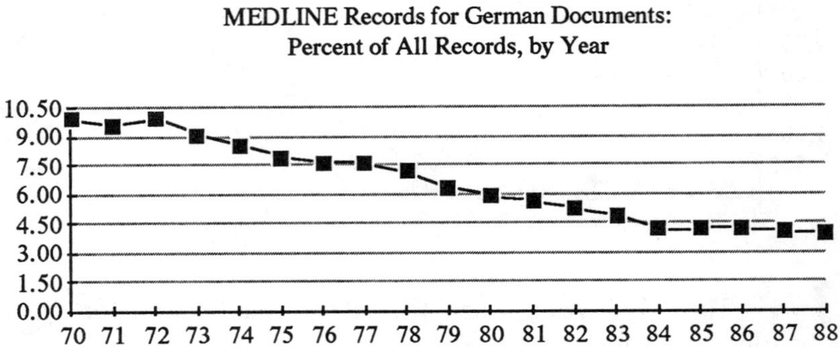

Spanish was stable across the years.

Danish and Dutch were too erratic to be placed into a category with confidence.

Countries of Publication Represented

MEDLINE also contains a more diverse representation of countries:

Table 4-6: MEDLINE Countries of Publication

Country of Publication	Number of Records
United States	2,195,498
England	752,086
Germany, West	449,022
Russia	388,432
Japan	271,929
France	243,952
Netherlands	228,213
Switzerland	173,048
Italy	166,059
Poland	107,665

Table 4-6, Continued

Country of Publication	No. of Records
Germany, East	106,834
Denmark	85,338
Czechoslovakia	77,406
Sweden	75,719
Canada	71,473
Australia	50,629
India	46,780
Belgium	41,963
Spain	41,894
Austria	27,420
Hungary	26,532
China	24,951
Norway	24,855
Romania	24,000
Brazil	23,853
Yugoslavia	21,630
South Africa	20,079
Israel	19,093
Mexico	16,560
Scotland	16,130
Finland	15,108
Bulgaria	14,776
Argentina	12,165
New Zealand	11,720
Chile	7,548
Ireland	7,460
Turkey	5,427
Taiwan	5,249
Thailand	4,943
Singapore	3,977
Egypt	3,766
Kenya	3,096
Greece	2,992
Portugal	2,780
Venezuela	2,556
Korea	2,263
Puerto Rico	1,952
Senegal	1,920
Tunisia	1,872

Table 4-6, Continued

Country of Publication	No. of Records
Zimbabwe	1,811
Malaysia	1,779
Philippines	1,402
Uruguay	1,284
Lebanon	1,269
Pakistan	1,258
Colombia	1,191
Morocco	1,177
Jamaica	1,077
Nigeria	973
Indonesia	899
Cuba	782
Sri Lanka	749
Peru	706
Hong Kong	635
Northern Ireland	610
Ethiopia	576
Papua New Guinea	575
Uganda	516
Zambia	495
Luxemburg	460
Iran	337
Bangladesh	334
Panama	289
Ecuador	251
Guatemala	222
Costa Rica	221
Madagascar	154
Iceland	144
Algeria	126
Syria	119
El Salvador	116
Ghana	115
Dominican Republic	112
Iraq	112
New Guinea	94
Paraguay	71
Malta	38
Wales	34
Jordan	20
Monaco	19

Table 4-6, Continued

Country of Publication	No. of Records
Barbados	18
Honduras	18
Bahamas	15
Haiti	8
Nicaragua	5
Bolivia	3
Virgin Islands	3
Arabia	2
Total	5,954,437

Representing materials from ninety-seven countries, the file is geographically diverse. A comparison of these data, gathered in June of 1989, with earlier figures from 1985, shows that the file itself increased by 13%. Some of the countries contributing smaller numbers of records increased dramatically (Panama, for example, by 605%, and Cuba by 41%) but it did not take a large number of records to generate these increases. Of the more substantially represented ones, higher than expected increases were found for the United States and England (both at 17%), Japan (at 15%), the Netherlands (at 21%), China (at 42%), and Austria (at 82%).

The top nine producers of materials were examined. They were the United States, England, France, Italy, Japan, the Netherlands, Russia, Switzerland, and West Germany. These countries contributed 2.5% or more to the file. These represented 80% of the file. In order to study 99% of the file, forty countries would have to be examined.

Country of Publication Trends

Representation of countries from which works were published showed a much more pronounced worldwide trend. Records from documents published in Russia decreased, although not substantially. Those from France were for the most part stable.

Countries receiving increased attention included Italy, Switzerland, and West Germany. More dramatic increases were exhibited by England, which nearly doubled the number of records present for each year of publication; Japan, which increased by about 50%; the Netherlands, which nearly tripled, and the United States, which increased its contribution just over twice its earlier figures. The pattern for Japan will serve as the example:

MEDLINE Records for Documents from Japan:
Records by Year

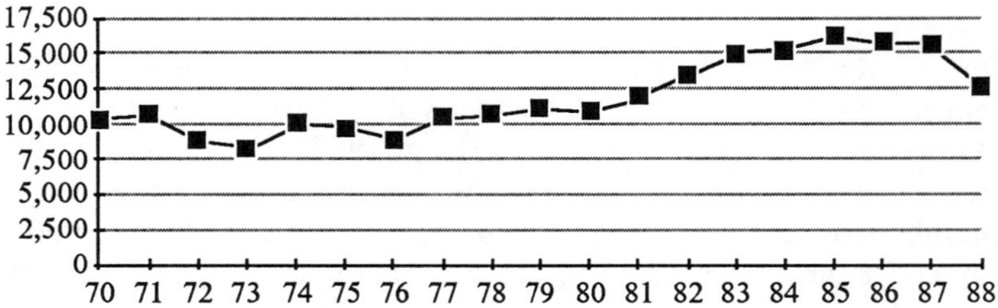

As a proportion of the database, the patterns for these countries were different. The pattern for France was stable. Records from Russia, Switzerland and West Germany decreased. And the decrease was dramatic: in 1970, these countries contributed 24% of the file, and, with the exception of France which decreased by 2%, maintained that proportion until the late 1970's. However, in 1987, their representation was just 16.5%.

Those countries increasing representation included England, Japan, the Netherlands, and the United States. The United States made the largest gains, from just over 30% to just under 45%:

MEDLINE Records for Documents from the United States:
Percent of All Records, by Year

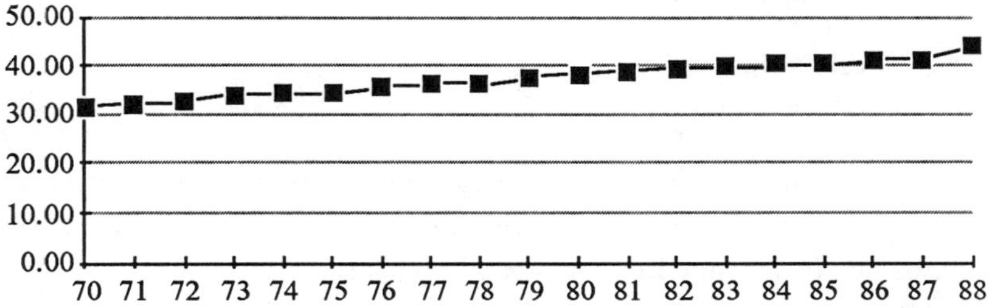

The three other countries combined only increased from 20% to 23% from 1970 to 1987.

The pattern for the proportion of documents from Italy was too erratic to confidently classify.

National Criminal Justice Reference Service (NCJRS)

The National Criminal Justice Reference Service was developed and managed by Aspen Systems in Rockville, Maryland, for the National Institute of Justice. It covers law enforcement, the courts, the criminal justice system, crime prevention and related topics.

Types of Documents

The file includes the following types of materials:

Document	35,828
Article	31,968
Microfiche	28,618
Book	4,402
Paperback	2,955

Film	1,415
Kit	382
Audio-cassette	288
Dissertation	92
Pamphlet	48
Filmstrip	36
News media	20

Not all records in the file have been categorized as a specific type of document, and examination of the records with no category yields no consistency in their type of records.

Languages Represented

The NCJRS file included the following languages, in the following frequencies:

Table 4-7: NCJRS Languages

Language	Number of Records
English	84,374
German	1,853
French	1,807
Dutch	409
Spanish	336
Italian	255
Swedish	180
Polish	101
Japanese	71
Danish	41
Finnish	30
Flemish	23
Slovene	22
Russian	20
Serbo-Croatian	20
Norwegian	18
Portuguese	18
Hebrew	12
Afrikaans	7
Czech	5
Ukrainian	5

Table 4-7, Continued

Language	Number of Records
Hungarian	2
Thai	2
Greek	1
Total	89,612

In the earlier study, the top nine languages were examined: English, Dutch, French, German, Italian, Japanese, Polish, Spanish, and Swedish. These represented 99% of all records. The updated data, however, showed that the dramatic increase in the file has been in records for English documents. Only the first five languages were required to cover 99% of the file. The remaining four have been updated, however, for comparative purposes.

Language Trends

The chart of all records for each year of publication graphically illustrates the dramatic growth, and decline, of the number of records added to the file each year.

NCJRS Records:
All Records, by Year of Publication

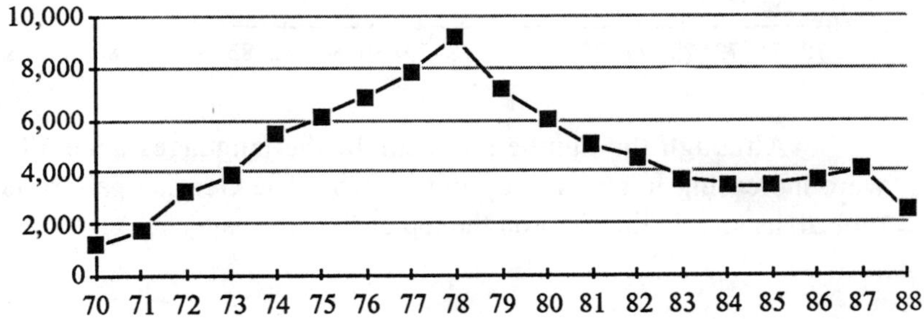

This chart showing the numbers of publications for all years of publication is radically different from all other files. In the others, the basic graph against which the language tables are compared generally is an upward slope, showing a growing database. Each year, a larger number of documents are added to the file. This database peaked in 1978, and declined through 1985 in numbers of records for each year of publication. A slight trend toward increasing the number of records added for each year may be seen for 1986-87.

These comments, therefore, have been based upon that pattern, rather than a growth pattern. Consistency was sought between the patterns exhibited by the languages, and the peak 1978 pattern shown in the chart for all records.

All of the languages examined declined after 1979, including English. The pattern for English materials most closely follows that of the entire file, primarily because English comprises such a substantial portion of the database:

NCJRS Records for English Documents:
Records by Year

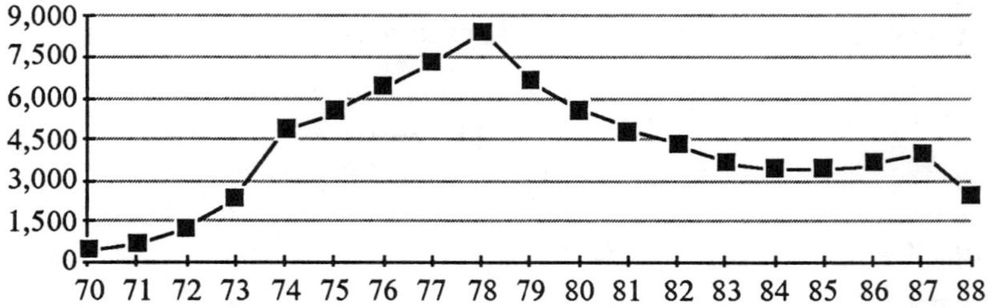

Although the numbers are small, the languages generally were increasing in strength up until 1978. The German pattern is typical, as seen in the chart on the top of the next page.

NCJRS Records for German Documents:
Records by Year

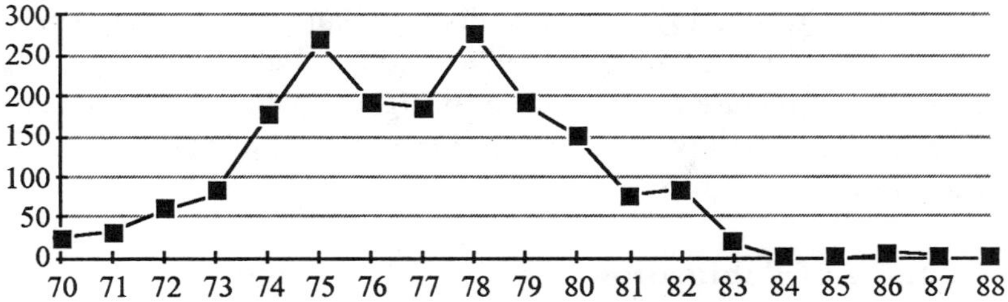

Polish and Japanese, however, were most prevalent in 1976 and declined from there, and Italian did so in 1977. Dutch managed to maintain its 1978 strength in 1979 but then decreased in numbers for each year.

The records as a percent of the file show a different representation of the activity of the file.

The English percentages showed a steady increase as a percentage of the database across the years. In other words, even as the database itself was in decline (post 1978), the English proportion was continuing to grow. After 1984, the proportion was consistently over 99%.

NCJRS Records for English Documents:
Percent of All Records, by Year

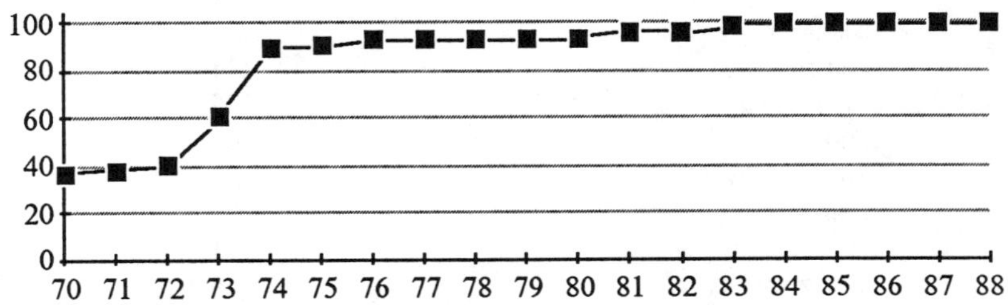

As a percentage of the file, all languages reached a year of strength in the 1970's and then simply declined. The declines began in different years, however: for Dutch, it was 1971; for Spanish, 1973; for French, 1974; for Swedish, Italian and German, 1975; and for Japanese and Polish, 1976. Notice that by 1977 (a year before the database itself reached its highest point) all of the non-English languages were decreasing in numbers of documents for each year.

Country of Publication

The file is international in the variety of countries from which its documents have been published:

Table 4-8: NCJRS Countries of Publication

Country	Number of Records
United States	72,830
United Kingdom	4,423
Canada	2,909
West Germany	1,878
Australia	1,606
France	1,071
Netherlands	806
United Nations	400
Sweden	377
India	373
Switzerland	348
Italy	280
Japan	271
Belgium	258
Poland	111
New Zealand	109
Finland	101
South Africa	94
Denmark	76
Israel	71
Venezuela	64
Norway	56
Ireland	52

Table 4-8, Continued

Country	Number of Records
Mexico	52
Spain	48
Hong Kong	43
Yugoslavia	43
Panama	39
Iran	34
Brazil	29
Costa Rica	26
Philippines	20
Austria	16
Hungary	16
Soviet Union	15
Singapore	11
Argentina	9
Colombia	9
Saudi Arabia	9
East Germany	8
Greece	7
Czechoslovakia	6
Papua New Guinea	5
Puerto Rico	5
Thailand	5
Antigua	4
Chile	4
Malaysia	4
Taiwan	4
Kenya	3
Nepal	3
Virgin Islands (US).	3
Bahamas	2
Bangladesh	2
Egypt	2
Indonesia	2
Jamaica	2
Nigeria	2
Pakistan	2
Peru	2
Portugal	2
Sri Lanka	2
Virgin Islands (British)	2
China	1

Table 4-8, Continued

Country	Number of Records
Iceland	1
Lebanon	1
Samoa	1
South Korea	1
Tanzania	1
Uruguay	1
Western Samoa	1
Total	88,279

This represents 88,279 records, from 72 different countries. The earlier study examined data for those countries which accounted for 99% of the records in the file: the United States, Australia, Belgium, Canada, France, India, Italy, Japan, the Netherlands, Poland, Sweden, Switzerland, United Kingdom, and West Germany. While the United Nations is not technically a country, it was included as a value in this file and was counted as a country for this analysis. For the updated data, these countries account for 96% of the file. The same countries will be used, however, for comparative purposes.

Country of Publication Trends

The patterns for materials from the United States mirror the file itself because they represent such a large (82%) proportion of the file.

Those for Australia, Canada, Japan, the Netherlands, Switzerland, and the United Kingdom are very roughly the same: they increased to 1978, and then declined in subsequent years. However, after 1982, the decline for some countries was much greater: for materials from Japan and Switzerland, the additions were negligible. Records for materials from West Germany followed the post-1978 pattern, but also reached nearly the 1978 level in 1975. The chart for Japan is illustrative and is shown at the top of the next page.

NCJRS Records for Documents from Japan:
Records by Year

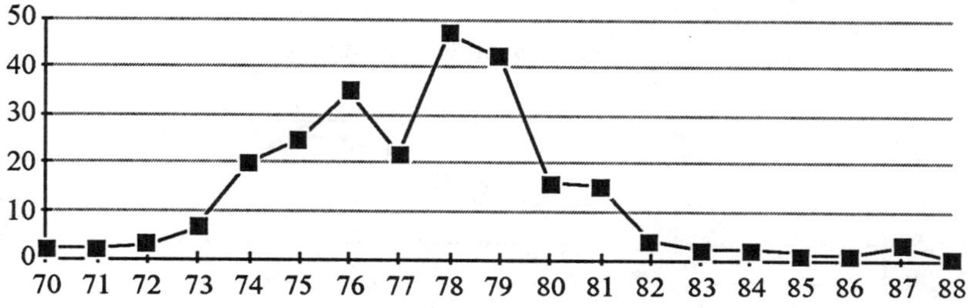

Other countries had peak years before 1978 (Belgium, 1974; Poland, 1976; Italy, 1975 and 1977) or after 1978 (Sweden, 1979; India, 1981). However these also showed the same post-1980 pattern of consistent decline in representation. Records for works from Italy, Belgium, Poland and France were virtually non-existent. Records designated as from the United Nations showed no consistent pattern.

The proportion of the database that these records represent continue the picture of a database declining in international activity.

The United States chart shows the continued dominance of that country over the file:

NCJRS Records for Documents from the United States:
Percent of All Records, by Year

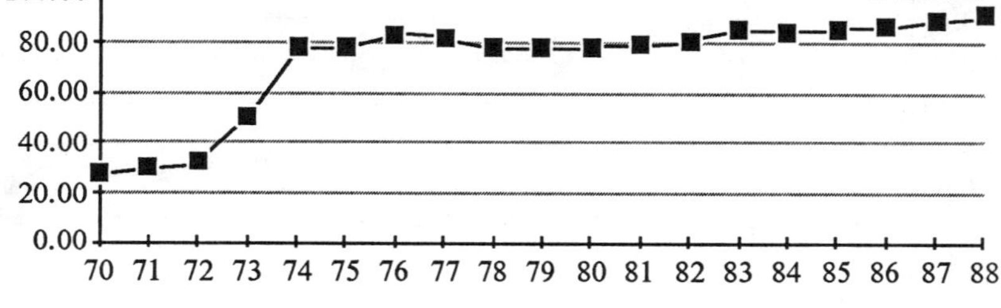

Only for Australia can a case be made that its strength is increasing in the database, but its percentage is very small:

NCJRS Records for Documents from Australia:
Percent of All Records, by Year

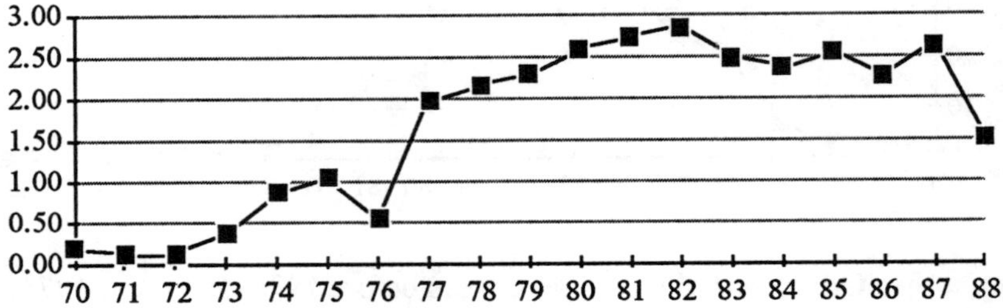

For records from Canada and those labeled "United Nations" a much weaker case can be made of increases, especially after 1985 for Canada and 1986 for the United Nations records.

For Japan, the Netherlands, the United Kingdom, and Sweden, there was promise of growth until 1978 or 1979. India continued to grow until 1981, as did Switzerland until 1983. Then various rates of decline set it, Switzerland's being dramatic and final. The example of Japan illustrates the pattern:

NCJRS Records for Documents from Japan:
Percent of All Records, by Year

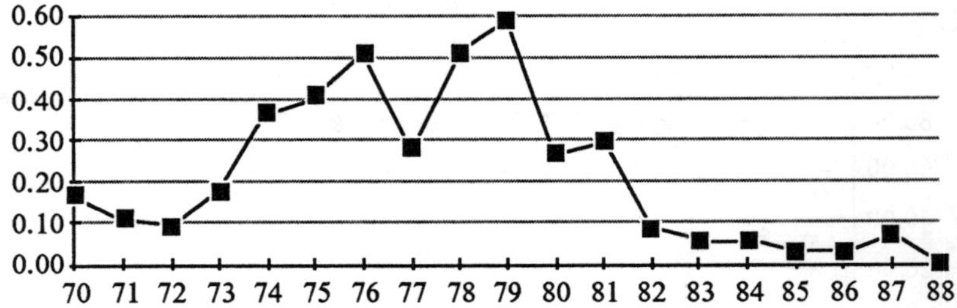

Even before the peak year of 1978, several countries were already declining in strength: Belgium's peak year was 1974, Poland's was 1976, and for West Germany, France, and Italy the top year was 1975. Decreasing interest in these countries was evident before major cut-backs began.

Overall, the data suggested that despite the decrease in the number of documents acquired overall, there was an increase in the number of materials added from some countries such as the United States, and an overall (that is, not connected to the decline point of 1978) decline in the acquisition or identification of materials from other countries.

Oceanic Abstracts

Oceanic Abstracts is a database concerning all issues and characteristics of oceans. It is produced by Cambridge Scientific Abstracts in Bethesda, Maryland.

Types of Documents

The database contained the following types of documents:

Journal Paper	16,714
Conference Paper	7,010
Reports	683
Book Chapter	526
Conference Proceedings	327
Books	254
Report Section	139

Since the data were originally gathered, these figures have increased dramatically. This is probably the result of both coding changes and increased numbers of materials such as journal articles.

Languages Represented

As the oceans are of international concern, we might have expected a large representation of languages. However, there were but 25:

Table 4-9: Oceanic Abstracts Languages

Language	Number of Records
English	181,259
French	4,385
Japanese	2,346
Spanish	1,349
Russian	1,172
German	1,060
Chinese	610
Italian	276
Norwegian	145
Portuguese	137
Korean	82
Persian	82
Polish	56
Serbo-Croatian	44
Multiple languages	31
Swedish	25
Arabic	21
Dutch	6
Hebrew	6
Danish	4
Icelandic	4
Slavic	4
Afrikaans	3
Finnish	1
Greek	1
Total records:	193,109

The major countries of the speakers of these languages all have a coastline. It would be interesting to have been able to determine whether there are very many records from land-locked countries, however it was unfortunately not possible to assess countries of publication. Also, the populations of many land-locked countries speak one of the above languages (such as countries in South America). Few countries can be set in a one-to-one relationship with a language. It can be pointed out however that Bulgarian, Hungarian, and Czechoslovakian are not on this list, but they do appear in the MEDLINE and BIOSIS lists.

In the earlier study, twelve languages out of twenty-two languages accounted for 99% of the entire file. In the updated data, 99% of the file can be accounted for with just English, French, German, Japanese, Spanish, and Russian. It is not the case that the non-English materials have been decreasing: Chinese increased from 118 to 610 records, and Japanese from 1,660 to 2,344. However, in other files even the less frequent languages showed some increase. In this database, only the more frequent languages showed substantial increases. In order to maintain the comparison, however, the full twelve languages have been brought up to date (the six mentioned above, plus Chinese, Italian, Korean, Norwegian, Persian, Polish, and Portuguese).

Language Trends

The language trends for this file were perhaps the most difficult to assess and describe of all the databases. There just were no clear trends in the vast majority of cases: most language patterns were just erratic. In other databases, the patterns showed an increasing number of records for each year of publication, or a decreasing number, no matter the overall picture for the database itself. Or, the number of records for each year followed the general pattern for the database itself. Even the general trend of the database itself, there was not a clear trend toward growth. It increased the number of records for each year from 1974 until the middle 1980's, and then decreased by nearly 25% by 1987.

Oceanic Abstracts Records for all Documents:
All Records, by Year of Publication

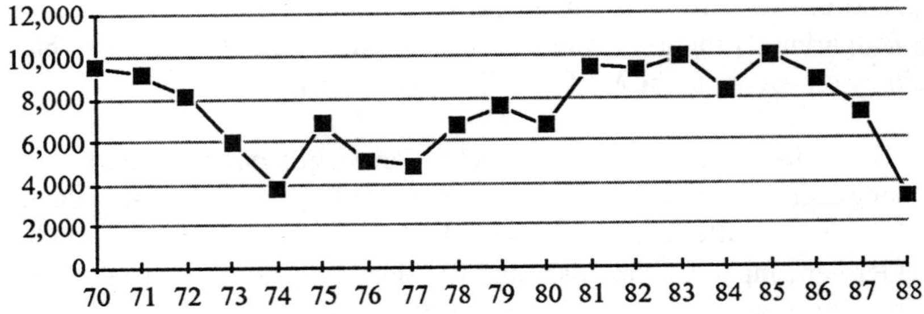

The only chart to mirror the database growth patterns was English, because it was such a large percent of the file, and its percentage remained high.

Oceanic Abstracts Records for English Documents:
Records by Year

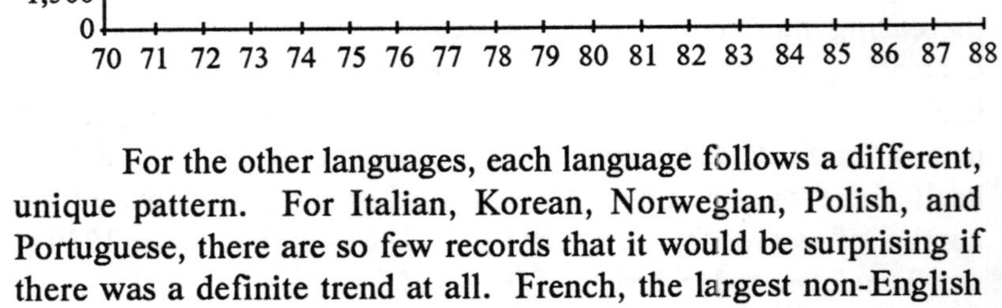

For the other languages, each language follows a different, unique pattern. For Italian, Korean, Norwegian, Polish, and Portuguese, there are so few records that it would be surprising if there was a definite trend at all. French, the largest non-English language, is typical in its erratic behavior, as seen in the chart on the top of the next page.

Oceanic Abstracts Records for French Documents:
Records by Year

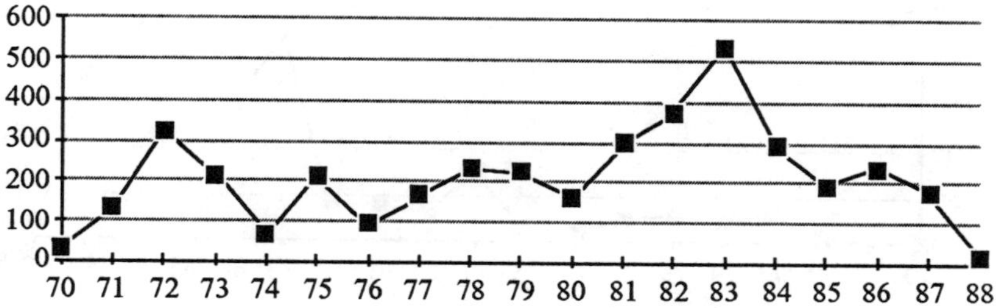

But it certainly did not show signs of increasing after 1983.

The only language to show potential for growth was Chinese. However, it was only recently (1983) in evidence in the file, and its total numbers are not very high.

Oceanic Abstracts Records for Chinese Documents:
Records by Year

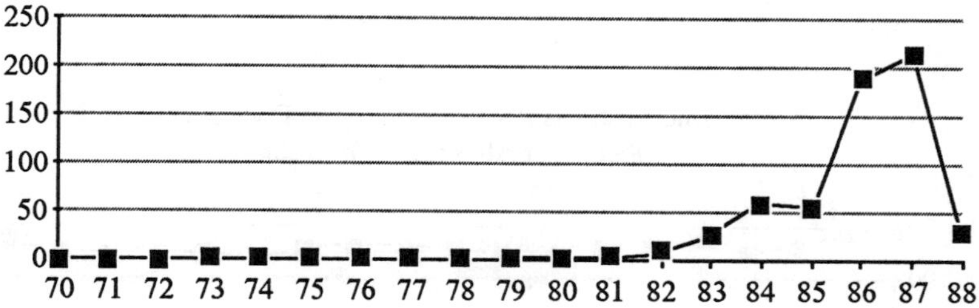

The pattern for Russian indicated a healthy increase in the middle 1980's, and 1987 was a strong year, in contrast to the remainder of the database. The chart for Russian is on the next page.

Oceanic Abstracts Records for Russian Documents:
Records by Year

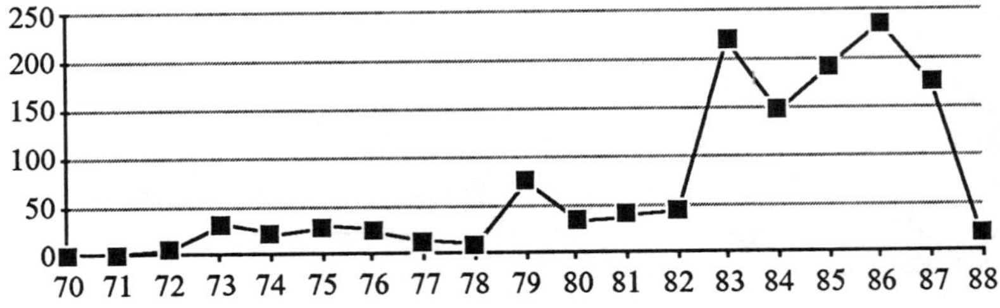

The records for Persian are in a category by themselves. When the actual records were examined, they were clearly in Portuguese, and not Persian. There appears to have been an error on the part of DIALOG or Oceanic Abstracts. The problem was identified by the database provider and reported to DIALOG. It has not been corrected at this writing.

The languages as a percentage of the database do not improve the picture of the file. English, of course, dominates the file.

Oceanic Abstracts Records for English Documents:
Percent of All Records, by Year

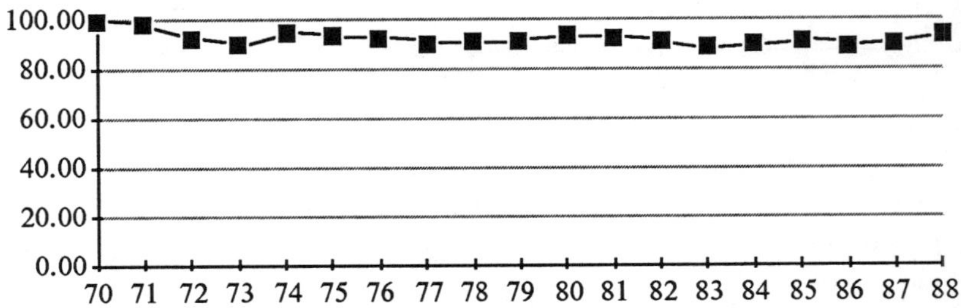

The proportions of Portuguese, Norwegian, Italian, and Polish are all less than one percent, and do not show a clear trend in any particular direction. Korean is also small, but shows growth for even 1988.

Of the more heavily represented languages, French, German, Spanish, and Japanese, no clear patterns emerge. Russian show some potential for continued and increased growth, as does Chinese. However, their raw numbers are not as optimistic, as noted above.

Summary Languages

The file originally contained a summary language field. It was removed from the file in 1986, after the initial research on the database was completed. The data in this section, therefore, reflect the file in 1986. It still is of interest to see the relationships between the major and summary languages present at that time. The following summary languages were represented:

Table 4-10: Oceanic Abstracts Summary Languages

Summary Language	Number of Records
English	4,530
French	1,680
Japanese	552
Russian	549
Spanish	465
German	289
Italian	122
Portuguese	54
Polish	32
Korean	27
Chinese	26
Swedish	17
Norwegian	15
Arabic	7
Dutch	2
Serbo-Croatian	2
Slavic	2
Romanian	2
Afrikaans	1
Total	8,374

Note that not all documents in non-English languages contain summaries. While it is of interest to simply note that English, French, and Japanese were the top three languages for summaries, of more interest were the patterns in which the languages were paired. Were French documents frequently summarized in English, for example, or in German?

For individual tables, the less used languages (Afrikaans, Romanian, Slavic, Serbo-Croatian, Dutch, Arabic, Norwegian, and Swedish) were ignored. The remainder of the languages made up 99% of all instances of summaries.

A word about the analysis of the summary languages is important. In the analysis process, each instance of a summary language was paired with the major language of its document. For each document, there may be several summary languages, and these were counted as separate instances.

Two languages were paired with only English abstracts: Icelandic, and Greek. The remainder of the document languages had abstracts in other languages as well.

Records in eleven languages indicated summaries only in their own language and English. These are presented below, with the percentage of all of the English summaries that they represent. For example, there were twenty-six documents in Chinese with Chinese summaries, and 109 Chinese documents with English summaries.

Table 4-11: Oceanic Abstracts English Summaries

Primary Language	Same Language	English Summary
Chinese	26	109
Norwegian	15	48
Dutch	1	2
Japanese	551	1013

Table 4-11, Continued

Primary Language	Same Language	English Summary
Korean	26	43
Polish	21	43
Slavic	2	3
Swedish	17	18
Afrikaans	1	1
Arabic	1	1
Serbo-Croatian	2	2

The remainder of the records for major languages included summaries in languages other than English and their own. These are shown in Table 4-12. The primary language is shown first, with the summary languages that were paired with it. For each summary language, the percent of all summaries for that primary language is given, followed by the actual number of summaries. For example, for the Danish primary language, half of its summaries were in English and half were in Dutch. There were only two summaries, one in each language.

Table 4-12: Oceanic Abstracts Major and Summary Languages

Primary Language	Summaries (%, #)
Danish	
	English (50, 1)
	Dutch (50, 1)
French	
	French (46.6, 1,597)
	English (37.5, 1,699)
	Spanish (1.4, 49)
	Italian (1.2, 42)
	German (.4, 13)
	Polish (.3, 11)
	Arabic (.2, 6)
	Russian (.1, 3)
	Romanian (.1, 2)
	Portuguese (0, 1)

Table 4-12, Continued

Primary Language	**Summaries (%, #)**
German	
	German (37.2, 1,062)
	English (48.7, 1,699)
	Russian (9.1, 65)
	French (4.8, 34)
	Italian (.1, 1)
	Spanish (.1, 1)
Italian	
	Italian (43.3, 78)
	English (46.7, 84)
	French (10, 18)
Portuguese	
	Portuguese (35.6, 53)
	English (60.4, 90)
	French (3.4, 5)
	Spanish (.7, 1)
Russian	
	Russian (55.1, 481)
	English (44, 384)
	German (.9, 8)
Spanish	
	Spanish (38, 414)
	English (59.3, 646)
	French (2.4, 26)
	German (.3, 3)
	Italian (.1, 1)

The file included values in the Summary Language field only beginning in 1973, and they were no longer included after 1986. These data were even more difficult to assess than the languages of documents because the language patterns were so generally erratic and there were so many years which do not show incidence of a given language at all.

Because the data are old and could not be updated, it is difficult to make firm statements about trends for the future use of summary languages. Clearly, Portuguese, Chinese, and Polish showed little promise at that time for becoming major summary languages. Japanese, Spanish, Korean, Italian, and German showed some promise in growth, but numerically they are all under 120 items each year. Russian and French were the most frequent non-English summaries used, and as expected, English was the dominant summary language for the file.

The percentage of summary languages that these languages represent was a little more interesting. First, Polish, Italian, and Chinese were not significant contributors. Spanish, German, Portuguese, and French did not increase, although Spanish and French in 1974 together made up 40% of all summaries recorded. If there were data for additional years, they might have shown that Russian, Korean, and Japanese continued to grow in influence. Interestingly, English decreased its share, from over 60% to less than 50%:

Oceanic Abstracts: Records for English Summaries,
Percent of Summaries Added, by Year

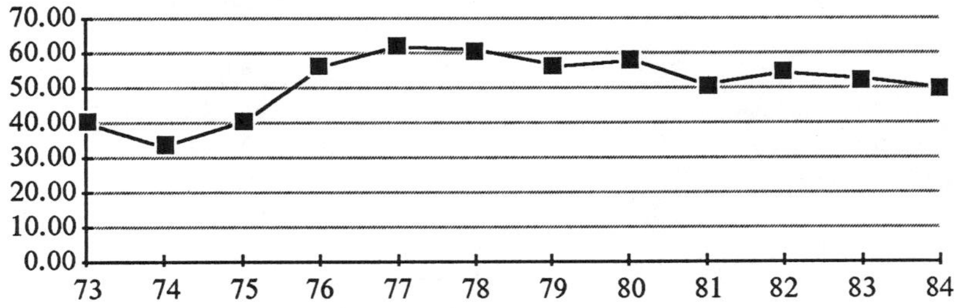

In other words, there was an increasing number of non-English language summaries being recorded in the early 1980's.

PAIS Information

The Public Affairs Information Service (PAIS) database focuses on economic and social issues. It is produced by Public Affairs Information Service, Inc., in New York City, from the collection of the New York Public Library.

Types of Documents

The database contained primarily periodical articles:

Periodical	178,664
Monograph	80,394
Special Issue	994

Languages Represented

It included materials in just six languages, as table 4-13 shows:

Table 4-13: PAIS Languages

Language	Number of Records
English	195,033
German	45,184
French	32,104
Spanish	14,410
Italian	10,375
Portuguese	5,711
Total	302,817

All languages have been evaluated because there were so few.

Since the earlier study, English has increased significantly, while the non-English materials have not. This is apparent in the charts for the file in the Appendix. An increasing number of records were present for each year of publication, with a decline in the last year. That decline was expected: as in other databases all 1988 processing most probably has not been completed.

The entire database has been decreasing in the number of records for each year of publication:

PAIS Records:
All Records, by Year of Publication

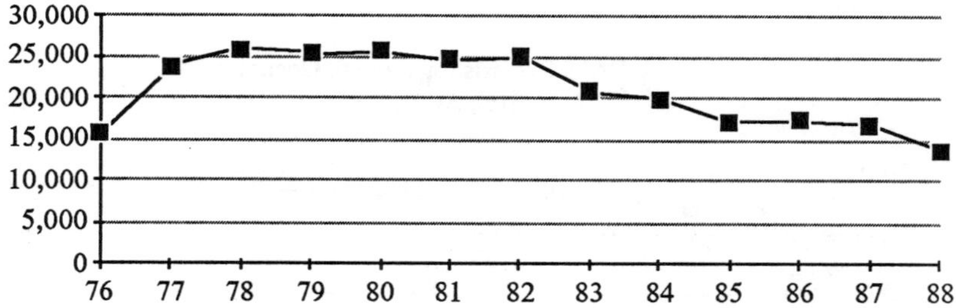

All of the languages except Spanish decreased in the number of records added for each year of publication. A typical example is Italian, in which records were present in the late 1970's in numbers around 700, but in the middle 1980's the figure had dropped to the 300 range.

French records also declined in number and percentage of the database. Its chart is shown at the top of the next page.

One result of this was an increasing proportion of English language records in the database. Even though the actual numbers did not increase, its proportion of the file did. The chart illustrating this is also on the next page. Spanish increased somewhat.

PAIS Records for French Documents:
Records by Year

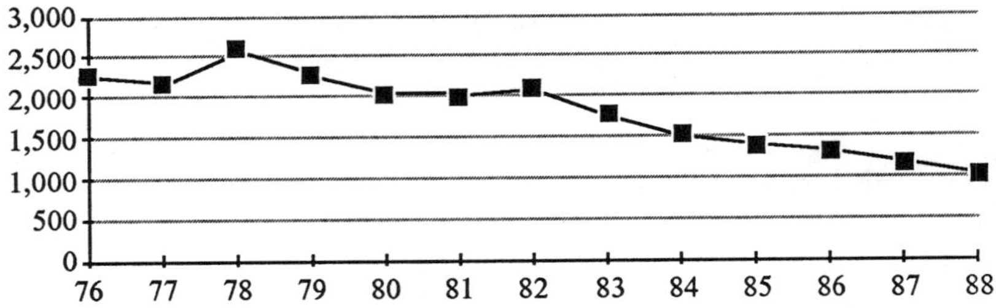

PAIS Records for English Documents:
Percent of All Records, by Year

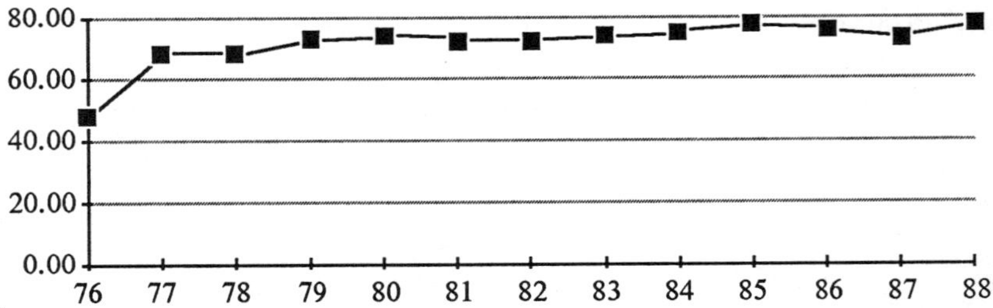

PsycInfo

PsycInfo, produced by the American Psychological Association in Arlington, Virginia, covers the world's literature in psychology.

Types of Documents

The file contains documents of a variety of types, as listed on the next page.

Journal article	407,702
Review	11,614
Book chapter	6,730
Dissertation	3,864
Other	2,555
Symposium/Convention papers	2,265
Technical report	1,359
Bibliography	330
Audio tape	1

Further examination of the database indicated that the remainder of records are journal articles, but it is not clear why some records are labeled as journal articles and and some are not.

Languages Represented

PsycInfo includes materials in 35 languages:

Table 4-14: PsycInfo Languages

Language	Number of Records
English	597,243
French	12,367
German	11,802
Russian	6,671
Spanish	5,987
XXXX[2]	5,433
Italian	3,175
Japanese	2,654
Polish	1,859
Dutch	1,584
Portuguese	1,583
Czech	1,521
Serbo-Croatian	955
Slovak	740
Swedish	664
Danish	621

[2]The XXXX value in the LA field was used before 1973, and it signified "non-English," according to PsycInfo staff. Before then, specific non-English languages were not identified.

Table 4-14, Continued

Language	Number of Records
Chinese	606
Hungarian	576
Romanian	557
Norwegian	420
Finnish	165
Hebrew	124
Yugoslavian	100
Slovene	98
Afrikaans	90
Catn (sic)	92
Bulgarian	84
Korean	46
Arabic	31
Flemish	12
Malaysian	10
Hindi	6
Slov (sic)	3
Ukrainian	3
Georgian	2
Lithuanian	2
Total	657,886

Like the other databases, the original data for PsycInfo were gathered in 1986. In the intervening years, some of the less-represented languages increased dramatically, but their numbers were still very small. Arabic, for example, went from two to thirty-one documents. Of the more heavily represented languages, French, Japanese, and Russian increased by 20-24%, while Spanish did so by 47% and Chinese by 44%.

Originally, twelve languages were studied: English, Czech, Dutch, French, German, Italian, Japanese, Polish, Portuguese, Russian, Serbo-Croatian, and Spanish, because they made up 99% of the database. The updated data, however, indicated that Portuguese, Dutch, and Serbo-Croatian could be omitted, because 99% was reached at the level of Czech. All twelve, however, have been updated for comparative purposes.

Language Trends

The growth of the PsycInfo file shows a roughly steady pattern of increasing numbers of records present in the file for each year:

PsycInfo Records:
All Records, by Year of Publication

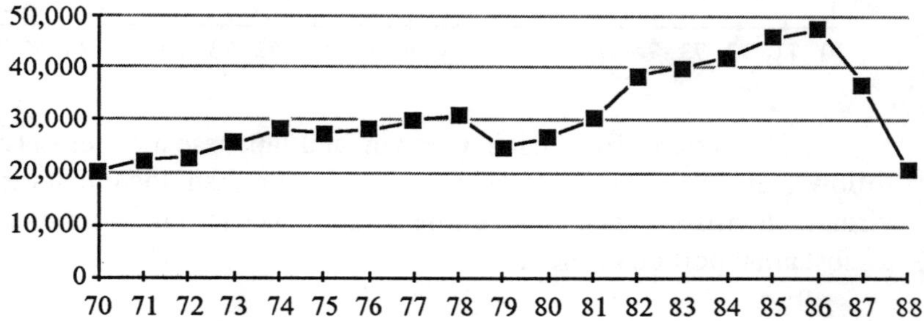

The patterns for Dutch, Italian, Portuguese, and Spanish showed an exaggerated version of the database trends, in that while they increased incidence in the 1980's, the increase was much higher than would have been predicted by the overall growth of the file. While the entire file doubled, the records for Spanish, for example, tripled (and quadrupled in one year) from their 1970's level. The chart for Spanish documents is at the top of the next page.

PsycInfo Records for Spanish Documents:
Records by Year

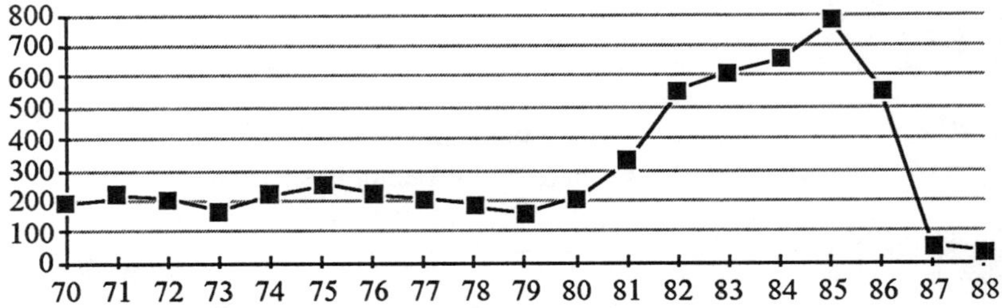

The patterns for French, German, and Japanese more closely follow that for the entire database. That for English documents is virtually a mirror image of the entire database because it forms such a substantial portion of the file.

The pattern for Russian was unique in that the numbers were in decline in the 1970's:

PsycInfo Records for Russian Documents:
Records by Year

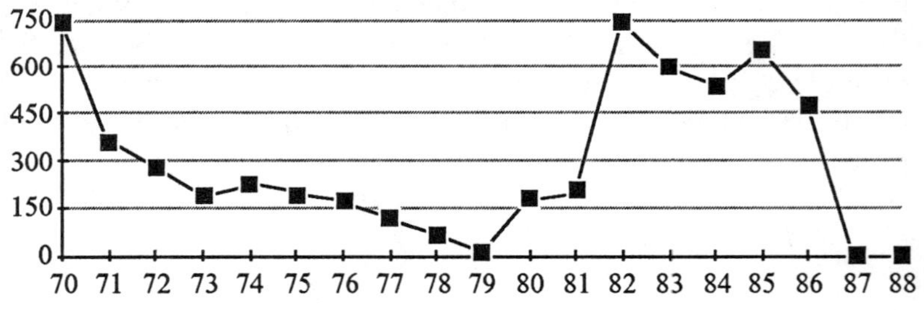

The patterns for Czech, Polish and Serbo-Croatian were erratic, and the records are not well-represented in the file.

When examined as a percentage of the database, the language patterns show a different picture of the development of the database.

English is clearly the dominant language of the file, consistently representing 85% of all records or more:

PsycInfo Records for English Documents:
Percent of All Records, by Year

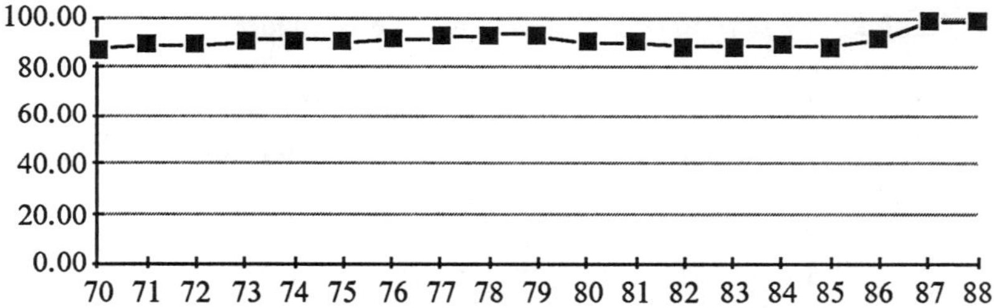

German and Czech show a gradual decline in the file, the fact being more significant for German because in the 1970's it made up more than 2% of the file. French, between 1.5-5% of the file, cannot be described as increasing as a portion of the file.

PsycInfo Records for French Documents:
Percent of All Records, by Year

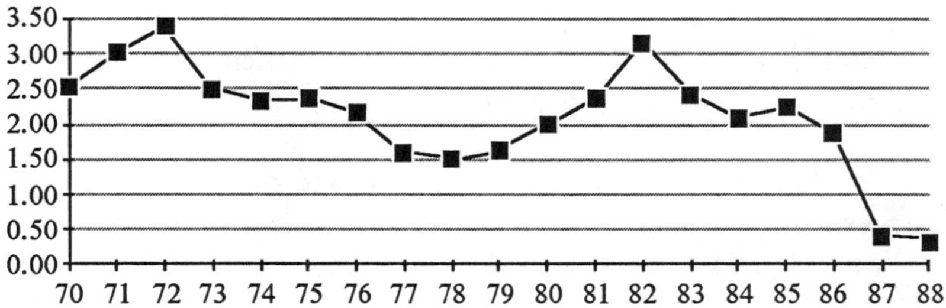

The patterns for Dutch, Portuguese, Russian, and Spanish roughly follow the pattern of the entire database. Italian does also, although with greater strength in the middle 1970's.

Japanese and Polish are roughly stable, although they are consistently around a half a percent of the file or less.

Summary

Overall, the number of languages and countries of publication in the eight databases is impressive. However, at this point there are several observations that can be made:

(a) With the exceptions of materials in Japanese and Chinese, databases are not increasing their content of non-English language materials.

(b) Languages producing fewer than 1,000 documents in these databases are virtually stagnant.

(c) Esperanto and Interlingua, which have been proposed as compromise languages for science, are making no headway whatever in these disciplines.

(d) Hindi and Arabic are widely spoken languages in the world; however, they are not languages of science.

(e) The one example of the cessation of the Summary Language field in Oceanic Abstracts indicates that at least for this discipline, there is decreased access to documents for scientists whose working language is not English.

Looking at the databases, their size and number of languages represented, we find the figures found at the top of the next page. (We are slightly comparing apples and oranges, because the number of records used is the number of records with the publication years 1970-88, whereas the languages represent all those in the database.)

Database	Number of Languages	Number of Records
Chemical Abs.	68	8,059,602
BIOSIS	54	4,103,138
GeoRef	53	1,110,093
MEDLINE	39	5,972,708
PsycInfo	35	657,886
Oceanic Abs.	25	193,109
NCJRS	24	89,612
PAIS	6	302,817

The size of the database does roughly correlate in several instances with the number of languages represented. The exception of course is PAIS, with only six languages. If size were indeed a factor, it would be expected to cover a number between Oceanic Abstracts and PsycInfo, or around thirty. NCJRS also is not consistent, being very small but covering a large number of languages.

CHAPTER 5

DATABASE PROVIDER POLICIES AND PRACTICES

Databases are developed by different types of organizations, with different policies and procedures for identifying, acquiring, and including materials in their databases. Some individual or group of individuals makes a conscious decision to include or exclude documents, by general policy or for specific documents. Some of these decisions for including documents are based on factors over which the agency or individuals have some control, others are not. These decisions to a large degree govern what is included in the databases described earlier.

The Justification for United States Databases to Include Non-United States Materials

One of the most dramatic reasons for inclusion of journals not published in the United States, or not in English, is when a major scientific advance is announced in an obscure journal. Just this happened in 1987, when K. Alex Müller of IBM's Zurich research laboratory published news of the development of a new warm superconductor in a German journal not frequently read in the United States (Rogers, 1987). The findings were so significant that many who read them did not believe them. But, subsequent findings proved them correct. Why exactly the scientist chose to publish outside the mainstream literature is not exactly clear, but the incident does point out the significance of some literature published outside of the United States, and not in English.

While it is pleasant and idealistic to consider the internationalism of science, the harsher realities of economic competition may be stronger driving forces in the development of services to provide access to the scientific and technical information of competitive countries such as Russia in the fifties and sixties and today from Japan. And with the competition recognized, the private sector organizations in the business of providing information also find economic reasons (i.e., profits) to try to meet this need.

There are several dimensions to the economic motivation. One is competition among secondary services themselves, and the efforts to capture sales of the services themselves to other countries. Another is a reaction to national economic conditions, and competition in the global industrial marketplace. Dr. Toshio Tsukahira, an Associate at Engineering Information, Inc., has stated, "This country can no longer afford the arrogance of indifference to scientific and technical developments in other nations, now serious competitors to our products in the area of domestic and international trade" (U.S. House. Availability... 1984).

John Woolston (1984) of the International Development Research Centre (IDRC) took issue with the idea of including foreign language materials to increase sales to developing countries. He saw the market for developed-country information products as very small. Prices were astronomical in those countries. There were no local information industry or capabilities of handling the information (such as obtaining the documents referenced) once it was obtained. He reported that sales of the National Research Council of Canada journals to developing countries were insignificant (10 titles, or 6.9% of sales), and quoted an anonymous United States publisher with sales to developing countries of 4.18%. He asserted that what the developing countries needed was access to *their own* information, and that there was no market for that here. However, he was not totally insensitive to developing country interests, and did favor cooperative information systems.

There is also the issue of the volume of production of research information. "Ten years ago the United States, with only five percent of the world's population, was generating about seventy-five percent of the world's technology. Now, the United States share is much less because of growing contributions made by countries like Japan. In other decades, our share may approximate only thirty percent" (D. Bruce Merrifield, U.S. House. Availability... 1984).

Various mandates had been issued to increase coverage of literatures beyond U.S. work. The Committee on Scientific and Technical Communication of the National Academy of Sciences and National Academy of Engineering (SATCOM) reported, for example, that "...while U.S. primary publications need cover only this country's research (though it may benefit our country to do somewhat more), abstracting in English must cover the worldwide literature in one unit if it is to serve the United States effectively" (National Academy of Sciences, 1969). Later the report mentioned the inclusion of foreign sources (but did not specifically deal with the issue of non-English language materials) as a special problem for expanding coverage.

The report further cataloged a variety of recommendations by various groups and conferences concerning the improved international exchange of information. Continually these acknowledged the need to work within the context of a world literature that is increasingly shared among many languages. Problems in doing so on an international scale included the uneven development of services among and within countries, language barriers, information policy restrictions, standards and management. The report indicated that progress at that time was satisfactory, and, since the U.S. stake was so high, it encouraged the continuation of efforts that were already underway. Action was to be taken at organizational, institutional, and governmental levels. While the direct and deliberate increase of foreign language materials is not an explicitly stated goal, such actions certainly fall within the scope of the variety of recommendations made.

One of the attitudes that must be taken into consideration in providing access to foreign information is the implicit admission that other significant work is being undertaken elsewhere. This can be difficult. Justin Bloom (U.S. House. Availability... 1984) suggests that a dual attitude is held toward Japan, for example. On the one hand, the country is viewed as imitative in its technological and research work. On the other, it is seen as a technological challenge to be met.

In order to be included in secondary services, there has to be a perceived need for the information by the database provider. And, there must be a perceived market for the information by the vendor if special hardware and software arrangements are to be made to handle non-English text. The most conclusive evidence of demand and use is the sale of a document. The National Technical Information Service (NTIS) reported in 1984 that Japanese documents overall sold 12 copies each, compared with an average of five copies for other foreign language materials (U.S. House. Availability... 1984). Generally, NTIS sales have reflected a growing interest in foreign literatures. In 1984, it reported a five-fold increase between 1981-83, without an increase in acquisitions. Note that NTIS is a self-supporting agency and cannot undertake unprofitable projects (Brady, 1986).

Low sales have had an effect on inclusion as well. NTIS has on occasion entered an English bibliographic entry for a Japanese document (apparently untranslated) into its database, but does so with great "reluctance," because such documents apparently seldom sell. Translations, at least for Japanese materials, are essential (U.S. House. Availability... 1984).

Whatever the source of the interest, demand from a target user group can certainly reveal a potential market for a service or modified service, and reveal the present level of access (or perceived level) to relevant information. Economic motivations have stimulated interest in Japan, and the establishment of the Japan Center at the Research Triangle Park in North Carolina further encouraged interest in the Asian region. A survey of faculty at the North Carolina State University in 1982 yielded results that indicated that over 30% of the respondents were interested in finding out more about Japanese activities in their fields, but that they had no way to do so (U.S. House. Availability... 1984). The survey was specifically to assess the demand on campus for "Japanese - related programs and services."

The field of magnetics has been particularly concerned about the influx of Japanese research, innovation, and market share. According to Clark Johnson, President of the Magnetics Society of the IEEE, over the last twenty years Japanese papers contributed to their International Conference on Magnetics had grown in number to the point where they were the majority in 1984. The concern is economic. "We in the United States are in imminent danger of losing both our technological leadership and a large portion of the computer peripheral disk drive business to the Japanese. The Japanese have already taken over virtually all of consumer electronics and a large fraction of semi-conductor manufacturing" (U.S. House. Availability... 1984).

Barriers to the Inclusion of Non-English Materials

The barriers to providing access to non-English information or information produced in other countries can include:

(1) Perception of the nature of research in the country. Japan, for example, may be viewed by some as a "technological imitator" (U.S. House. Availability... 1984), thus producing less original, and less significant, research.

(2) Availability of linguistically trained staff to handle the materials. Language training has been declining in school systems in the United States for a number of years, and languages such as Japanese are just not as popular as French or Spanish, and are much more difficult to learn.

(3) The structure of scientific and technical information production within a country. In Japan, for example, it is more common to publish in journals of the researcher's university than in refereed journals of a learned society, for example, of which there

generally are fewer. And, the development of secondary services there has not occurred rapidly within the country (U.S. House. Availability... 1984). In Latin America, there is a tendency to publish outside the country entirely for disciplines such as physics and chemistry (Rosenberg and Whitney, 1986).

(4) Copyright and other laws regarding access and dissemination. In Japan, for example, all government reports are copyrighted, and releases for re-publication must be secured from the author, sponsor, and issuing agency (U.S. House. Availability... 1984). Further, the country's copyright laws allow distribution of the *Japan Report: Science and Technology* only to the United States government offices, and not to its private sector.

In its attempt to provide access to Japanese literature, *Engineering Index* identified four "risks" or chief areas of difficulty:

(1) acquiring the materials themselves,

(2) the shortage of translators,

(3) the cost to develop and provide the services, and

(4) a general lack of funds.

Landau's statement at a Congressional Hearing suggests that some titles might not even be available for distribution outside the country for political reasons. It is interesting to note that he did not include cooperative efforts with any Japanese organizations as a means of either providing access "locally" (e.g., using Japanese translators rather than English-speaking ones) or as a means of cutting costs. Nonetheless, the costs of improving access can be indeed considerable (U.S. House. Availability... 1984).

Apparently, *Engineering Index* attempted to convince large American companies to finance an abstracting service to cover a thousand Japanese engineering journals, and failed (Brady, 1986).

A further factor influencing inclusion can be the format and layout of the journal article itself. Japanese articles, for example, apparently seldom carried abstracts, thus it would be impossible to simply have translated an abstract into English (U.S. House. Availability... 1984). The need to translate the entire article, or to read it and prepare an abstract would add substantially to a budget for such a service.

The acquisition of scientific and technical information is of national concern for other countries, for example Japan (U.S. House. Availability... 1984). In 1957, the Science and Technology Agency of the Prime Minister's Office formed the Japan Information Center of Science and Technology (JICST), which continues information activities today. Its mission was specifically to acquire foreign scientific and technical information. While several other proposals were put forward to establish other programs to strengthen the effort, the chief center for such activities remains the JICST. The agency acquired materials, produced abstracts written by scientists and engineers, maintained an online information system, and accessed foreign databases.

Secondary Services' Policies and Procedures[1]

A variety of factors specific to the administration of a secondary service will affect the nature of materials included and the languages given coverage. These include the staff's language skills and the ability to tap other languages, the financial ability to acquire the materials themselves, user expressions of desire for foreign language materials, and even the position of the parent organization in the field.

[1]This section is not intended as a catalog of current (1989) database policies and practices. Those will affect records that are added to the database later this year and next. Rather, it describes policies in place during the period studied. It is these policies that affected the distribution patterns documented in this work. The information was gathered by discussions with staff of the various organizations and from examination of informal documents of the organizations. If no data are presented for a given file, the staff were unable to offer explanations for trends found or further details about the database.

The National Academy of Sciences report notes that *Psychological Abstracts* (at that time, 1969) processed domestic materials by abstracting them immediately. Foreign materials "typically are peripheral rather than core sources" and included so that readers would be aware of them, but abstracted only when relevance was established or a "body of related articles had accumulated."

In his 1981 study, Victor Rosenberg was told by editors of *Sociological Abstracts* that the users of the service would prefer that foreign language materials not be included in the bibliography, because they were not used anyway and were just clutter.

The development of the *Population Bibliography* between 1970 and 1979 provided a good example of organizational influences and the development of a database. The database was developed from the holdings of the Technical Information Service Library of the Carolina Population Center. The Center itself was one of the pre-eminent research and training institutes in the field and was known internationally for its work. Visitors from other countries were common, both for short as well as extensive visits. The library staff were able to get to know these researchers and policy makers, and were able to solicit documents from them representative of their own research and that of their countries. Upon their return home, many would continue contributing to the collection.

The Center had been funded to a large degree with funds from the U.S. Agency for International Development. When foreign aid funding generally was sharply curtailed, the Center virtually ceased all international activities, and the library itself ceased acquiring materials from other countries in the numbers to which it had become accustomed. The bibliography consequently ceased to be as strong as it had been in its coverage of foreign materials.

Chemical Abstracts, as another example, did not acquire any materials from the People's Republic of China for six years ending in 1973, due to "an interruption of scientific publication in the People's Republic of China." The service resumed coverage with 28 journals in 1975.

Chemical Abstracts

Chemical Abstracts (CA) intends to cover the world literature in chemistry, broadly defined. Its coverage includes, for example, nuclear physics and biotechnology. It is a part of the American Chemical Society (ACS) in that the Society owns the service, but it is self-supporting by user fees.

Chemical Abstracts covered 12-14,000 journal titles, during the period of the study, the vast majority of which were provided to CA gratis. A small number were purchased, some were obtained on an exchange basis. The majority were not abstracted cover to cover. Relevant articles or chapters were chosen from each document.

Chemical Abstracts staff did not feel that resource costs affected the ability of CA to cover the literature. The majority of materials were provided gratis, therefore the rising costs of scientific serials did not affect its ability to cover the field. Financial resources could be considered a slight factor in the coverage of the field, in that CA might not have obtained a difficult to acquire, expensive conference proceeding that contained only one or two relevant articles out of a far greater number. He felt that the stature of the abstracting service, its circulation, and use was sufficient to convince publishers that it was of benefit to them to be included. Libraries purchase journals as demanded by users (within financial limits), and if journals are frequently cited in a secondary service, they are requested by users.

Staff felt that the coverage of the field was essentially complete, and that an influx of additional resources would not significantly enable the staff to cover the field (in contrast to other database providers, as described below.) It would only enable the staff to cover borderline materials.

Mathematical Reviews

Mathematical Reviews is one of the premier secondary services for the field of mathematics in the world. It was started in the 1930's in Germany by Otto Neugebauer as an attempt to cope with the flood of literature in the field, at that time amounting to 2,000 papers each year (Boas, 1979). The journal was originally called *Zentralblatt für Mathematik,* and became *Mathematical Reviews,* published under the auspices of the American Mathematical Society when its editor left Nazi Germany for the United States and Brown University. Its first issue appeared in 1940. The journal moved to Ann Arbor, Michigan, in 1965 to be closer to consultants at the University's Mathematics Department. After several moves in the town, it currently occupies a converted brewery/ice cream factory in (ironically enough) a German neighborhood. The journal currently covers 40,000 papers a year (Shackman 1985).

Documents to be included in *Mathematical Reviews* were passed through the library first. Generally, two copies of materials were requested from providers: one for the library, and one for the reviewer. If two could be provided, a copy was microfilmed for the library collection. A special arrangement existed with the University of Michigan library, in that for 300 journals, it relied on the University Library for its backup copy.

The documents to be sent to reviewers were examined by ten editors, with language and subject specialties. While they had special languages of concern, all of the editors had to be able to handle Russian. The editors selected the reviewers to which the documents would be sent.

At *Mathematical Reviews,* a written policy dictated that article titles were translated into English, as were analytic titles in books and collection titles. The original titles were given for Russian, German, and French titles, as well as the English titles. The policy for collection titles was a recent change: earlier policies dictated that collection titles were translated into English.

In the early 1980's several major policy changes regarding the language of reviews were put into effect: before this time, reviewers could write in English, French, German, or Italian. In 1980, Italian was dropped. In 1984/85, all other languages except English were dropped as review languages. This policy change was made because it was determined by the Society that English had become the language of mathematics, and that mathematicians around the world were capable of working in English. Also, there were additional costs of typesetting and managing the foreign language text that could be avoided with an all-English journal.

Summary languages were given only if the summary languages were different from the main language.

English was considered the "understood language" in that if no language was given, the item or summary was in English. Otherwise, the language was noted.

English, French, and German were considered the important languages for mathematics in the early 1980's. However, the publication works currently only in English.

National Criminal Justice Reference Service (NCJRS)

The NCJRS file was designed to serve the criminal justice and law enforcement community. It "was established in 1972 by the National Institute of Justice to meet the information needs of the Nation's law enforcement, criminal justice, juvenile justice, investigative, and correctional communities. Since that time the mission ... has been enlarged to include victim assistance, statistics, and dispute resolution."

The Service offered standard information center services of reference, referral, interlibrary loan, and selective dissemination of information. Of particular interest was the *International Summaries* publication in which each issue contained a four-page English summary of recent non-English criminal justice literature. The documents resulting from the service were held for the collection for a time and then were turned over to the Library of Congress.

The database has been developed on a contract basis with Aspen Computer Systems, for the National Institute of Justice; the result of the acquired materials is the Service itself. Since 1975, the Institute has granted Aspen the contract, which is renegotiated and renewed every three years.

The stated policy of the Service as to acquire materials "on the basis of their utility to the criminal justice community ... determined primarily by their subject, length, and currency ... document type, and completeness. ... The criteria are reviewed regularly to ensure responsiveness to the research initiatives of the National Institute of Justice and to the users' information needs." Documents must have been at least half-concerned with a topic on a long series of subjects, such as civil rights, crime prevention and deterrence, and criminal investigation. Included on the list was comparative research on criminal justice in foreign countries.

In 1978, the Institute required that the Service add 1,000 items to the database each month. The Institute also mandated an emphasis on foreign materials. At that time the Service had translators on staff, and on call. During that period also the International Summaries series was started, with two issues per year.

In 1981, a new contract was issued. President Reagan had just taken office and the federal government was under massive budget cuts. The Institute required that the number of documents be halved to 500 per month, and the Service decided that the foreign language and non-United States materials could be most easily cut. They were the most expensive to acquire and process. Journal subscriptions were also cut. A second way to cut down on the number of documents added was by year of publication: under earlier contracts, materials published in the last five years were candidates for inclusion. Under this contract, materials published in the last three years were candidates.

In 1984, another new contract was issued. The number of documents to be added each month was halved again. The Institute and the Service worked out formal selection criteria, and again the foreign language and non-United States materials were dropped in priority. The emphasis for the Service shifted to materials and information to be used by workers in the field (policemen, judges, corrections officers, for example) rather than on that used by researchers, so the policy could be justified. In earlier days of the file, the emphasis was on information for researchers.

In negotiations for the 1987 contract, the Service recommended to the Institute that the number of documents be increased, and that the number from other countries and in other languages be increased.

These policy shifts are dramatically evident in the language trends for the file described in Chapter 4 and in Appendix 3.

In part, because the database had been developed by a private-sector company, there as a sense in working with the staff that its procedures were proprietary. Nonetheless, some comments can be made about them. The Service maintained an International Document Exchange program with fifty members (in the early years of the database). More recently the group numbered about forty-three members, who were primarily Ministries of Justice and research organizations. These contacts generated about a hundred documents a month, and the staff is expanding the program.

Materials were sought primarily from the European countries, because their law enforcement techniques and practices were considered to be as advanced, or more advanced in some cases, than those used in the United States. Latin American countries' practices are considered to be less advanced than those in the United States. A question on terrorism yielded the response that materials on this subject are collected, but they are generated in the United States and in Europe, not in the Middle East.

Staff at the Service indicated that they had had no user complaints about the database cuts, either from the perspective of the size of the file or from the perspective of materials from other countries or in non-English languages. The staff also indicated that there were no complaints about the *inclusion* of non-English publications.

Oceanic Abstracts

Oceanic Abstracts intends to cover the world literature on issues and characteristics of the oceans.

The database began in 1964 as the *Oceans Citation Index*. It was bought by Data Courier in 1981, and subsequently by Cambridge Scientific Abstracts again in 1981. The acquisition by Cambridge of the Aquatic Sciences and Fisheries Abstracts in 1982 meant the acquisition of several scientific centers which served as contributors, in France, Germany, the United Kingdom, the Soviet Union and other countries. The acquisition was a general takeover, for commercial interests, in areas of common interest. The input centers are government organizations, and the file is a unique collaboration between the government centers and the commercial Cambridge enterprise.

The database service monitored the publications of the field, and received primary publications in house. It was a depository for documents from the National Oceanographic and Atmospheric Administration and Sea Grant publications. After the materials are processed, they were not kept (for example, for document delivery or supply, or for local retrieval).

The service did not purchase any materials for inclusion in the file. Rather, it relied on materials from the input centers, and by the identification of articles of relevance by copying tables of contents of journals at university libraries. The Editor chooses those articles which will be included in the database from these tables of contents, and relevant articles are then obtained as photocopies. The service paid for this copying.

Decisions on which articles are included were made based on relevance, rather than any other criteria. However, the service was limited to 1,500 citations per bi-monthly update period. Decisions were never made on the basis of language, rather, staff tried to cover the entire field no matter what language was being used. Staff felt that the file could be doubled in size given the amount of material that was being published, but there were not the personnel and administrative resources available to handle the work load.

The service had a cutoff date of two years. Given delays in acquisition of materials from other countries, this could have a slight effect on the acquisition of materials from other countries due to mail delays. However, it should not be significant in this case because the Service was relying on in-country input centers, which should have more ready access to local materials.

Indexing capabilities are maintained in-house for the major European languages (English, French, German, Spanish and Italian).

An abstract was prepared in-house if the document warrants it (according to decisions made in-house). An indicative abstract was prepared in house; if an informative abstract was required, outside resources are used.

Russian titles were transliterated if there was no English provided. The abstract is not translated.

The database has received only compliments about the foreign language materials. The service estimated that the *Oceanic Abstracts* users are primarily from the United States, while only half of the Aquatics users are located in the country

Public Affairs Information Service (PAIS)

The *Public Affairs Information Service (PAIS)* file on DIALOG is a combination of the English language printed service, *PAIS*, and the *Foreign Language Index (FLI).* The *FLI* was developed as a corollary service to *PAIS* in 1971, in response to perceived market demand. Both strive to cover the social sciences, emphasizing public policy. The *FLI* file was added to the DIALOG file in 1976.

The resources for both files were those of the New York Public Library, and especially the Economic Affairs Division . The Indexers and Editors scanned all journals as they were received, and included relevant articles from any journal. They reviewed the General Research Division's periodicals. And, they participated in decisions on the acquisitions of serials.

The FLI staff acquired a half a dozen journals directly (while PAIS acquires "lots"), and there was no effort to expand the number of journals acquired, primarily because of resources available. The selection policy therefore reflected that of the relevant divisions of the NYPL, in subject scope and type of materials (newspapers, weeklies, and polemical journals are omitted).

Languages covered were limited by the staff's abilities, and to those languages considered the most important in the United States and Europe.

The staff of two individuals indicated that they had more materials available than they could handle, given the resources of the NYPL. They indicated, however, that most of the other materials that could be included would be relatively worthless. In other words, while the coverage could be greatly expanded given additional resources, the utility of the index would not be increased.

The staff (for *FLI*) consists of two professionals and five clericals, the latter of which are are shared with *PAIS*.

In 1985 *PAIS* began adding abstracts to articles and items included in the service, which further decreased the number of materials that could be handled. Without more staff, and the addition of abstracts, the staff could not keep up with the earlier number of documents that were listed. While the listings of those materials that were included was improved with this action, the sheer number of documents covered could not be increased.

The staff indicated that the chief factors that affected the development of the database were the staff available, and the acquisitions policies and practices of the NYPL. Events outside of the United States affected the inclusion of materials only indirectly. For example, when Argentina's government became democratic, the country began publishing in great numbers, and materials were simply more available for everyone.

The staff also alluded to general problems with DIALOG itself, which could affect the findings of this study. Apparently DIALOG has loaded files twice, which would double the number of publications for a given period of time in the languages under study. Staff indicated that she thought that these problems had been resolved.

MEDLINE

Discussions with staff at the National Library of Medicine did not yield explanations for why the database was exhibiting the patterns found. It was simply assumed that these are the patterns of the world's medical literature as a whole.

In 1980, Craig and Strain reviewed a 1977 collection development policy revision at the library that suggested what might have actually happened. In this revision, English language titles were given top priority for cataloging, and foreign language titles were given lower priority. The accompanying chart in the article showed a dramatic drop in the number of titles cataloged into the CATLINE database after 1977. In fact, between 1966 and 1968 more foreign language titles were being cataloged than English language titles. In 1973, half as many non-English titles were cataloged as English. In 1977, after the policy revision, the ratio was closer to seven to two. While the number of non-English titles *may* have been decreasing up to this point, the policy revision certainly speeded up the process. Admittedly, the CATLINE file covers books, while the MEDLINE file covers journal articles.

The rationale for the policy change was also telling: "Reduction in the number of foreign language titles cataloged was related to a decrease in the number of catalogers with expertise in the less widely known foreign languages." To restate the justification: it wasn't because of the literature, it was because of staffing resources.

The serials collection at NLM has also been growing far faster than the set of titles included in *Index Medicus.* The 1986 Long Range Report (National Library of Medicine, 1986) reported that while the number of journals covered by *Index Medicus* had increased 20%, the number of titles in the NLM collection had increased by 69% (from 13,888 to 23,470). This document was a series of reports by panels of experts which were adopted by the Board of Regents as the NLM long range plan.

The experts noted several times the international responsibilities of the NLM, with statements such as:

"...we are increasingly dependent upon the results of foreign research, so we need data from other countries and we need to assure the availability of those data."

"The world of biomedical information is not limited to what is produced in the United States; it spans the globe."

"Although a considerable portion of the research done abroad is published and available to United States professionals through electronic databases and published literature, there is still a great deal that is only available in more local databases accessible through foreign national systems and networks." (NLM. Report of Panel 1).

Further, the report urges that more scientific methods be used to determine the journal titles included in *Index Medicus,* rather than "ad hoc judgements, opinions of consultants, and recommendations by subject and geographic experts." Given the difficulties in doing this kind of analysis, NLM faces a challenge in implementing this recommendation.

While the statements on international activities are to be praised, the reality of the budget for additional resources needed to carry out the recommendations told a different story. International activities (the "promotion of international exchange of biomedical information") were accorded the smallest budget allocation of $10,000. It did not appear to be as high a priority as the text would suggest.

Certainly it would not be impossible for NLM to build a collection that is more international and yet less concerned with foreign language materials. And certainly it would be possible to assert that the best literature is being included. This is just additional evidence to support the suspicion that *some* of the world's literature may be being omitted.

CHAPTER 6

DOCUMENT PRODUCTION DATA FROM PRIMARY SOURCES

The preceding chapters have presented a picture of a non-English scientific and technical literature that, according to the secondary services, is declining in numbers and frequencies except in certain cases, such as Japanese and Chinese. This chapter presents evidence that suggests that the incidence of non-English and non-US materials is increasing, rather than decreasing as the data from the secondary services show.

Scientific Literature Production Worldwide

While this study focused on literature that is visible through secondary services, it is useful to look also at estimates of literature production from different countries and in different languages.

A very early study in 1962 by Charles Bourne (Bourne, 1962) suggested that indeed abstracting and indexing services cover only "a fraction" of the world's scientific literature, and proposed that that fraction might be 25-30% of the literature. And, this fraction may be all of the significant work, while the remainder is "rehashes, popularized writings, and trivial contributions." At the time of his writing, MEDLARS was not yet available for searching biomedical literature, and the membership of the National Federation of Science Abstracting and Indexing Services (now the NFAIS) covered only 16,000 of the estimated 30,000 scientific and technical journals published in the world. If indeed the fraction of literature covered by abstracting and indexing services is perceived as the "significant" work in a given area, the secondary services are exerting a powerful influence through their judgements on what literature is sufficiently significant for inclusion. Bourne details the difficulties in assessing the amount of literature published, and noted the inconsistencies and unreliability of the data he presented.

He cites English as the most "predominant" language, accounting for half of the available literature, French and German are next, with Russian increasing to their level.

For country of publication, again, the United States is the most prolific, followed by Germany, France, and the United Kingdom. Russia may have been "more prominent" in chemistry as opposed to other subject areas. His "knowledgeable sources" suggest that Japan may be ranking third, and may be even higher for chemistry, biology, medicine, and agriculture. He cites a 1961 Library of Congress study (Bourne, 1962) which offers the following breakdown of journals published in various countries:

Country	Published Journals
United States	5,950
Germany (East and West)	2,950
France	2,660
United Kingdom	2,090
Italy	1,520
South America	1,500
Japan	1,470
USSR	1,450
Belgium	1,250
Red China	650
Spain	300
Korea (North and South)	150
Taiwan	150

He noted a CIA study which reported 2,295 science and technology journals from the USSR, and a National Science Foundation study which reported a total of about half that. Clearly it is difficult to ascertain the number of journals published in that country: and this suggests the difficulties worldwide in assessing scientific literature production.

Assessing production by language is equally difficult. A National Science Foundation study in 1960 estimated that English accounted for 50% of the total scientific and technical literature, while another 33% was in Russian, Chinese, and Japanese. The latter assertions were derived from the assumption that the literature published in the USSR, China, and Japan are in those languages respectively. A Unesco study three years earlier indicated that English accounted for 60% of "publications," but was not specific about what was included in the term "publications." Russian and German were set at 11% each, French at 9%, Japanese at 3%, Spanish at 2%, and others at 4%. These statistics are presented to show relative relationships, and not specific percentages. And, they are estimates and probably not very accurate, according to the cited report.

Bourne provided statistics given to him by the National Library of Medicine for 1958 language distribution of medical literature:

Language	Percent of Total
English	37.3
German	13.0
French	12.9
Japanese	7.8
Italian	7.5
Spanish	6.2
(Polylingual)	3.3
Russian	3.1
Other	8.9

He also presented data from NLM regarding the country of origin of periodicals received from 85 different countries, as shown on the next page.

Country	Percent of Total
USA	24.5
Germany	11.7
France	10.2
Japan	8.9
Italy	7.8
England	7.4
USSR	3.2
Other	26.3

Bourne further presented similar statistics for *Chemical Abstracts* for 1959. Dale Baker, President of Chemical Abstracts Service, reported to Bourne that fifty-two different languages were represented in *Chemical Abstracts,* and the most frequent were these:

Language	Percent of Total
English	50.5
Russian	16.8
German	9.7
Japanese	6.1
French	5.5
Italian	3.6
Other	7.8

Baker reported that ninety-seven countries contributed publications to the *Chemical Abstracts* service and that the most frequently represented were these:

Country	Percent of Total
USA	27.1
USSR	19.1
British Commonwealth	13.8
Germany	7.8
Japan	7.8
France	5.0
Italy	3.2
Czechoslovakia	2.0

Country	*Percent of Total*
Poland	1.5
Switzerland	1.3
Hungary	1.2
Sweden	1.1
The Netherlands	0.9
Others	8.2

Note that the sudden mention of the British Commonwealth means that this table could not be directly compared with "England" in other tables. This is just another of the difficulties in trying to locate reliable, comparable data.

Comparing the country of publication USSR at 19.1% and the language Russian at 16% suggests that some literature being published from the USSR is not being published in Russian.

Bourne also presents similar data for biology, compiled from the Biological Abstracts service:

Language	**Percent of Total**
English	39.0
French	13.0
German	13.0
Spanish	6.0
Italian	3.0
Slavic Tongues (several)	8.0
Others	18.0

He is also able to provide country of publication information from *Biological Abstracts,* as shown on the next page.

Country	Percent of Total
USA	26.75
USSR	11.40
Japan	5.85
Great Britain	5.83
Germany	4.22
France	4.10
Italy	3.15
Brazil	2.64
India	2.60
Canada	2.13

The 1985 *Science Indicators*, published by the National Science Board, addressed some of these issues in its overview of the progress of United States science each year. The company Computer Horizons processed tapes from the Institute for Scientific Information's Science Citation Index, which collected and presented lists of references and bibliographies in articles in 2,100 (1973) to 3,500 (1981) journals. The resulting citation index shows "who was citing whom" in different scientific disciplines.

Appendix B from the document presents the "U.S. share of world scientific and technical articles" for a variety of subject areas. Overall, all areas covered show a decline in the U.S. contribution, except for biomedicine, which shows a one percent increase. The overall percentages indicate a decline from 38% in 1973 to 35% in 1982. Those areas of interest for this paper are as follows:

Discipline	1973	1981	1982
Clinical medicine	43%	41%	41%
Biomedicine	39%	39%	40%
Biology	46%	37%	38%
Chemistry	23%	20%	21%
Earth and space science	47%	42%	42%

Notice that this study covered journal literature and does not describe the grey literature, or unpublished reports and papers, and does not not describe book literature.

The Unesco Statistics

Each year, the Unesco *Statistical Yearbook* includes a table presenting the number of books published by reporting countries, in their various languages. This is an important set of statistics because the numbers are reported by the countries themselves, rather than gathered by a secondary service. In theory and practice, a secondary service can (and does) introduce biases (positively and negatively) into the total counts of publications issued by a country or in a given language. Most, for example, include "the best" or "most relevant literature" in their fields. Further, it is traditional to think of books as more formalized interpretations and synthesized discussions of the state of a field, whereas journal articles (the focus of databases) are more current reports of scientific activity.

There are a variety of problems in working with the statistics. For example, there was no common definition of a "book." The statistics did not cover all of the countries in the world, only those that reported data to Unesco. The totals do not add up: the 1984 data for Zimbabwe, for example, list 193 titles published, of which 155 are in English, which is identified as the national language, but no other statistics are given for the languages of the remainder of the titles, even in the "Other" column. Unesco omits figures for the United States: they are simply not included. The tables also omit the United Kingdom and Germany. The tables therefore represent a display of the data that were reported to Unesco, and there is no way to know how this relates to the actual publishing universe.

The notes to the tables indicate additional difficulties in trying to obtain numbers which are comparable across national boundaries. For example, for the tables in the 1986 *Yearbook*, the following apply:

(a) Ghana: 312 titles are in English, 38 are in other national languages (the table separates figures by national language and selected other languages, but not all countries have one national language), and it does not say what the languages are.

(b) Mali: Data refer only to textbooks. No explanation is given for the treatment of textbooks for other countries.

(c) Reunion: Data refer only to first editions. No explanation is given for the treatment of other editions by other countries.

(d) Japan: Data do not include pamphlets. The table does not explain what a pamphlet is, nor what is done with other countries' pamphlets.

(e) Singapore: The data do not include government publications. No explanation is given for government documents of other countries in their entries.

Each annual statistical yearbook presents data for the latest available year for each country. The 1985 yearbook presents data for 1983 from Algeria, and 1981 from Ethiopia. The 1986 yearbook presents 1984 data for both countries.

There is nothing that could be done about the problems with the data, given the scope of this project. Contacting the countries for their definitions of books, and obtaining more details about what is included in each set of numbers, were beyond the scope of this work. These are the numbers that are publicly available and will be analyzed with these caveats in mind.

It is possible to begin to cope with the problem of missing data by making high, low, and medium estimates of what might have been published, to fill in the missing information. To show the degree of information that is missing, the following table lists the countries and the number of years missing information for each language. A value of "15" means that all years are missing, a value of "none" means none are missing and there are statistics for each year present. Note that "missing" means both that the country actually did not publish in that language and that the country might have been expected to publish in that language but for a year or years did not report any data.

Table 6-1: Unesco Countries Reporting Data

Country	Eng.	Fre.	Ger.	Russ.	Span.
Afghanistan	14	15	15	15	15
Albania	10	11	11	11	10
Algeria	13	13	15	14	15
Am. Samoa	12	15	15	15	15
Argentina	3	6	8	11	6
Austria	1	1	1	3	4
Australia	15	15	15	14	13
Bahrain	13	15	10	15	13
Bangladesh	13	15	15	15	15
Barbados	13	15	15	15	14
Belgium	4	4	3	9	6
Benin	15	14	15	15	15
Bhutan	15	15	15	15	15
Bolivia	13	15	15	15	13
Botswana	9	15	15	15	15
Brazil	10	11	11	14	12
Br. Virg. Is.	13	15	15	15	15
Brunei	8	15	15	15	15
Bulgaria	1	none	1	none	3
Burma	13	15	15	15	15
Byelor. SSR	12	12	13	4	13
Cameroon	12	12	15	15	15
Canada	7	7	7	10	9
Chile	5	5	11	15	5
Colombia	10	12	14	15	12
Congo	15	14	15	15	15
Costa Rica	14	15	15	15	10
Cuba	7	8	12	10	7
Cyprus	5	5	5	10	15
Czechoslovakia	1	none	none	none	3
Denmark	none	none	none	7	9
Domin. Repub.	15	15	15	15	12
Ecuador	14	15	15	15	14
Egypt	7	7	15	15	15
El Salvador	15	15	15	15	14
Ethiopia	12	15	15	15	15
Faero Is.	15	15	15	15	15
Fiji	14	15	15	15	15
Finland	none	none	none	none	9

Table 6-1, Continued

Country	Eng.	Fre.	Ger.	Russ.	Span.
Fr. Guiana	15	13	15	15	15
Fr. Polynesia	15	12	15	15	14
Gambia	14	15	15	15	15
Germany, DR	15	15	15	15	15
Germany, FR	15	15	15	15	15
Ghana	6	9	15	15	15
Gibraltar	15	15	15	15	15
Gilbert Is.	15	15	15	15	15
Greece	5	6	6	14	15
Grenada	14	15	15	15	15
Guam	13	15	15	15	15
Guatemala	14	15	15	15	14
Guyana	8	14	15	15	13
Holy See	6	6	6	15	9
Hong Kong	4	9	8	13	11
Hungary	1	1	1	1	5
Iceland	15	15	15	15	14
India	2	15	15	14	15
Indonesia	14	15	15	15	15
Iran	13	14	15	14	15
Iraq	12	14	14	15	15
Ireland	4	11	14	15	15
Israel	6	13	13	13	15
Italy	1	none	5	15	4
Ivory Coast	14	12	15	15	15
Jamaica	12	15	15	15	15
Japan	5	5	8	6	9
Jordan	11	15	15	15	15
Kenya	11	12	15	15	15
Khmer Repub.	15	14	15	15	15
Korea	4	4	6	14	14
Kuwait	11	13	15	15	15
Laos	14	14	15	15	15
Lebanon	14	14	15	15	15
Lesotho	14	15	15	15	15
Libya	13	13	15	15	15
Luxembourg	9	5	5	15	14
Madagascar	6	4	15	14	14
Malawi	10	14	15	15	15
Malaysia	2	11	13	15	15

Table 6-1, Continued

Country	Eng.	Fre.	Ger.	Russ.	Span.
Maldives	15	15	15	15	15
Mali	15	13	14	15	15
Malta	4	11	15	15	14
Martinique	15	14	15	15	15
Mauritius	1	2	14	15	15
Mexico	12	11	15	15	8
Monaco	15	11	15	15	15
Mongolia	15	15	15	15	15
Morocco	15	14	15	15	15
Mozambique	15	15	15	15	15
Nauru	15	15	15	15	15
Netherlands	2	2	2	9	8
Neth. Antilles	13	15	15	15	14
New Caledonia	13	12	15	15	15
New Zealand	6	13	14	15	15
Niger	13	12	15	15	15
Nigeria	13	15	15	15	15
Niue Is.	15	15	15	15	15
Norfolk Is.	14	15	15	15	15
Norway	3	3	3	15	13
Pacific Is.	13	15	15	15	15
Pakistan	7	14	15	15	15
Panama	10	15	15	15	15
Peru	4	9	10	14	5
Philippines	0	14	15	15	13
Poland	1	none	none	none	3
Qatar	12	14	14	15	15
Reunion	15	12	15	15	15
Romania	5	5	5	15	15
St. Christopher	14	15	15	15	15
St. Kitts	14	15	15	15	15
St. Lucia	13	15	15	15	15
Samoa	13	15	15	15	15
San Marino	15	15	15	15	15
Saudi Arabia	15	15	15	15	15
Senegal	15	11	14	15	15
Seychelles	13	14	15	15	15
Sierra Leone	14	15	15	15	13
Singapore	1	13	12	15	15
S. Africa	12	14	13	15	14

Table 6-1, Continued

Country	Eng.	Fre.	Ger.	Russ.	Span.
Spain	3	4	4	15	4
Sri Lanka	3	9	10	15	14
Sudan	12	14	15	15	15
Sweden	none	none	none	3	3
Switzerland	1	1	1	15	15
Syria	14	15	15	15	15
Tanzania	11	14	15	15	15
Thailand	2	7	14	15	15
Tonga	14	15	15	15	15
Trinidad & Tob.	13	15	15	15	15
Tunisia	12	11	15	15	15
Turkey	11	11	11	15	15
Uganda	14	15	15	15	15
Ukrain. SSR	6	4	5	4	8
USSR	3	2	2	1	4
Un. Arab. Em.	14	12	14	15	15
Upper Volta	15	14	15	15	15
Uruguay	10	13	15	14	12
Venezuela	14	13	14	15	14
Viet Nam	10	10	15	11	12
West. Samoa	13	14	15	15	15
Yugoslavia	none	2	none	none	15
Zaire	15	13	15	15	15
Zambia	13	15	15	15	15
Zimbabwe	11	15	15	15	15

Those countries which appear to have no data at all reported books published in languages other than these five. Of chief concern in all of this was the number of countries that generally do publish in large numbers, but reported their totals erratically (such as Canada and Japan). There were no figures contributed by the United States, the United Kingdom, and France: three of the largest publishing countries in the world.

High, Medium and Low Estimates

First, if a country had been submitting statistics fairly regularly, there was no good reason to presume that when data were missing, the country had simply not published anything at all that year. Therefore, it can be assumed that the country must have published something and would continue to do so in the same languages if statistics had been reported for those languages. Second, there was no good reason to presume that the missing years would be any worse, or any better, than the worst or best years reported.

Therefore, those countries reporting English titles for all but three years or less, or all but four years or less for French, German, Russian and Spanish, were isolated. For the low and high estimates, the missing data were filled in for each country with the lowest and highest numbers reported, respectively. For the medium estimates, an average of the lowest and highest year was taken, and the missing years were filled in with this average. The low estimates and actual figures are presented in this chapter below, all of the estimates are presented in Appendix 4.

The resulting graphs do not show the declining patterns that were frequently seen for non-English publications in the charts from the databases. If the patterns were not relatively stable, then they are showed an increase.

Admittedly this is a comparison between books (however they are defined) and periodicals (which make up the bulk of the databases). However, there is no good reason for a country to increase its book production and dramatically decrease its periodicals. A valid criticism would be that the books represent popular works, whereas the periodicals in the databases represent scientific works and more was indeed being published in English in science.

English

The number of countries reporting books in English declined from just over 50 in 1979 to just under 20 in 1984. The decline is quite clear during this period. There should be a dramatic jump in 1980, considering the number of countries that are reporting. And, the number of documents reported was more severely down than would be expected, given the rate of decline of the number of countries reporting.

Unesco Statistics: All Countries,
Number of Documents Reported in English

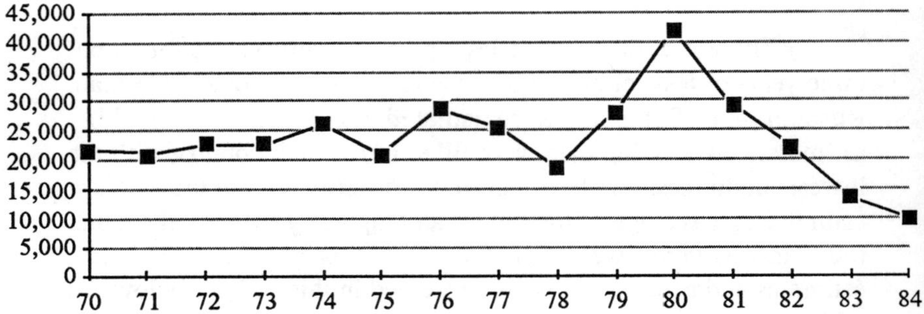

The three graphs of low, medium and high estimates show that if the countries had been reporting regularly, the figures would have been rising.

Unesco English Document Statistics:
Missing Data Filled in by Low Estimates

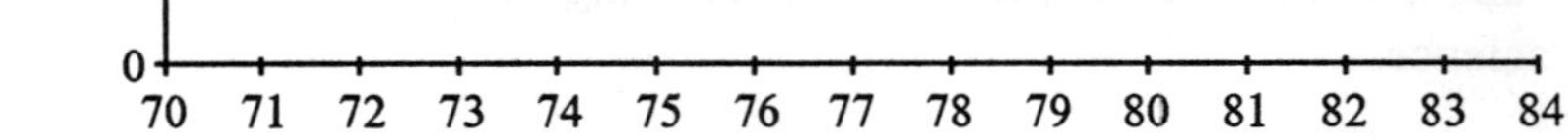

The figures for the chart above include those countries for which reporting was fairly regular. It omits a sizeable number that reported irregularly. The point of the exercise is to identify trends, not necessarily exact estimates of production.

French

The statistics roughly resembles the number of countries reporting data, with peaks in 1976 and a rise in 1970-1972, and a decline in 1980-83. Note, however, that the total number reported is higher than would be expected according to the number of countries reporting, in 1973, 1978, 1979, and 1984.

Unesco Statistics: All Countries,
Number of Documents Reported in French

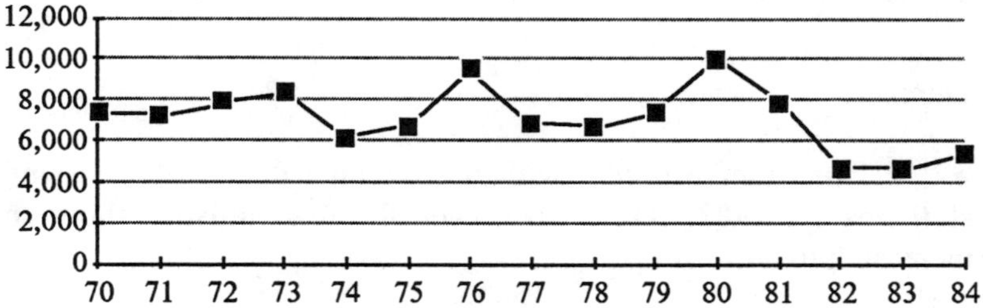

The actual number of documents reported is down over the period of study, but the charts of estimates suggest that this is because there are missing years of data. Adding in the estimates for missing years stabilizes the data and provides a graph pattern that is definitely not decreasing. The medium and high estimates in fact show an increasing pattern. The low estimate is less clear, but could be interpreted as a very slight increase. The chart showing the low estimate is on the next page.

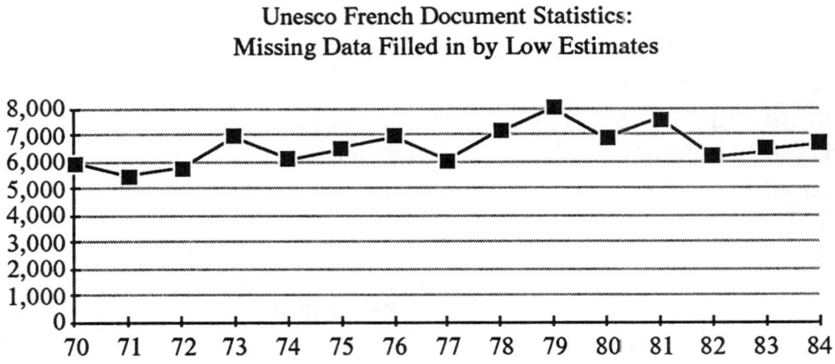

German

The chief problem with the German data is the missing statistics for 1983. Otherwise, even the raw statistics show an increasing trend.

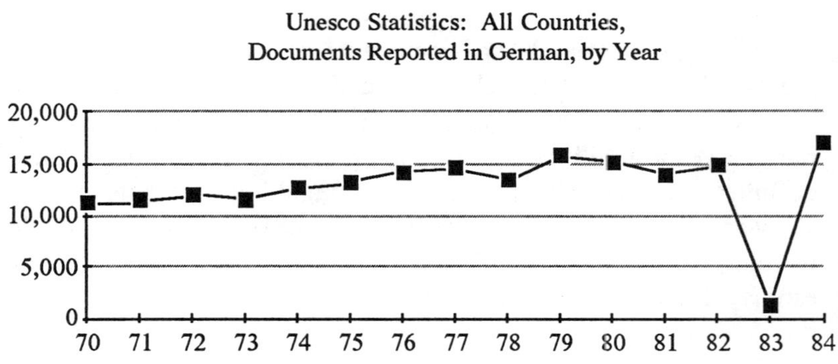

The number of countries reporting for 1983, however, was in contradiction to this pattern. Data for Austria and Switzerland were simply not reported. Data for the Germanies were also absent. The numbers are definitely higher than would be expected given the number of countries reporting in 1984.

Filling in the data with estimates simply confirms the trend, even with the low estimates.

Unesco German Document Statistics:
Missing Data Filled in by Low Estimates

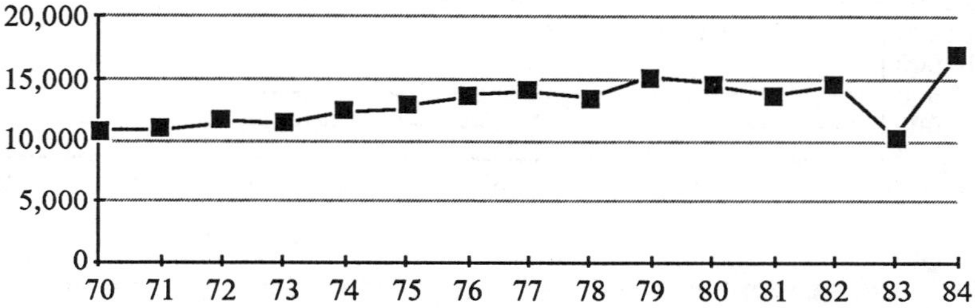

Russian

The chief problem with the Russian data is the missing statistics for 1974. The USSR simply did not report for that year. Ignoring that year, the trend for Russian materials was decidedly not down, and is up (comparing the last half of the period to the first) by over 10%.

Unesco Statistics: All Countries Reporting,
Documents Reported in Russian, by Year

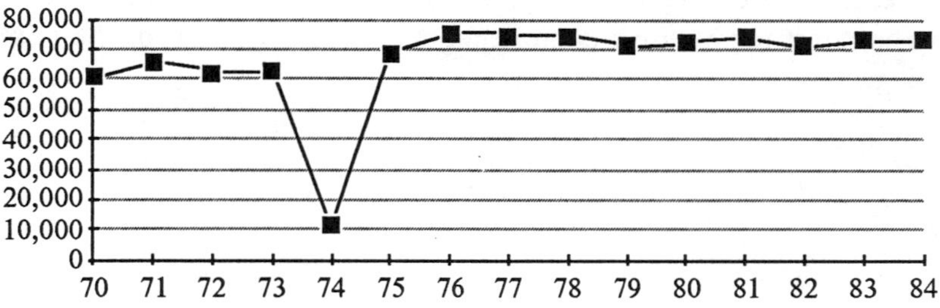

The low estimate chart simply confirmed the trend, which is clearer in the charts for medium and high estimates.

Unesco Russian Document Statistics:
Missing Data Filled in by Low Estimates

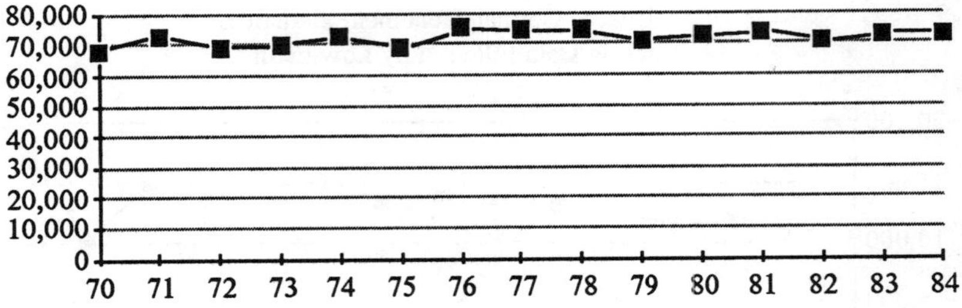

The number of countries reporting books in Russian has roughly increased over the period, from ten to 17 (in 1983) or 14 (in 1984). This percentage is far higher than the percentage increase of documents themselves.

Spanish

The number of countries not reporting data suggests that it was definitely missing data, rather than a decline in actual numbers, that accounts for the trends shown. The data for Spain are simply missing for 1978, 1980, 1981 and 1983, and those for the USSR are missing for 1978, 1979 and 1980. These are the largest producers of Spanish books according to the Unesco statistics, and these omissions account for the very low numbers in these years. The chart for Spanish documents is shown at the top of the next page.

Unesco Statistics: All Countries,
Documents Reported in Spanish, by Year

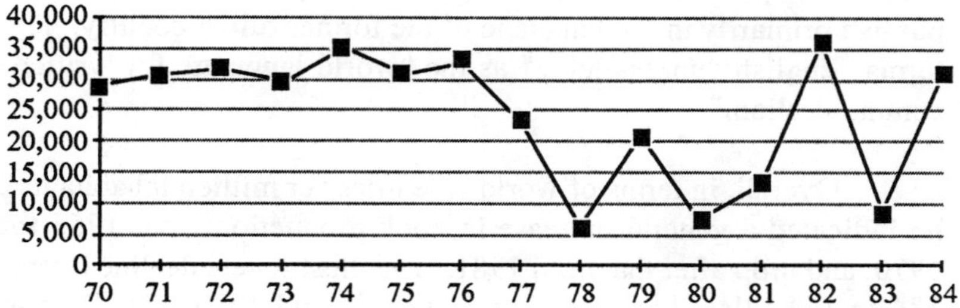

The Spanish records are highly erratic during the last half of the period of study. The data for those countries reporting all but four years or less suggest an upward trend. The insertion of low estimates suggests an uncertain pattern, the medium and high suggest a rising trend.

Unesco Spanish Document Statistics:
Missing Data Filled in by Low Estimates

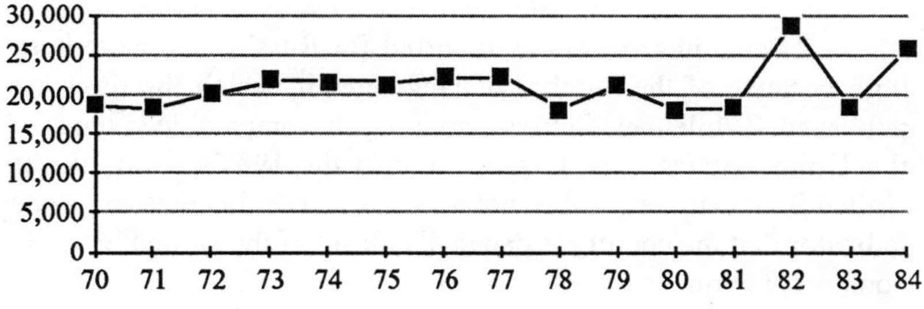

The Curwen Analyses

In 1986, Peter Curwen, a leading expert in the book industry, reported on a study of the industry (Curwen, 1986). He also described the inconsistent measures used to report book production, such as volumes versus titles, and translations of titles for countries publishing in multiple languages. He attributes the dominance of a few languages to colonization, as others have done before, and pointed out that except for India, the former colonies publish primarily in the language of the former ruling country. He terms English "unassailable" as the "world language for written communication."

Overall, in terms of worldwide titles per million inhabitants, he indicated a general increase in book production from 1955 to 1970, and drop after that until 1981. The most severe decline (from 184 to 164 titles) happened between 1975 and 1980. He did not offer reasons for this drop. The developed countries, on the other hand, saw an increase from 1970 to 1975 (420 to 431), and another jump to 500 titles in 1980. The only explanation for the worldwide drop appears to be a decrease in titles from Asia. All other regions of the world increased from 1975-80.

In his report, he commented on the Unesco statistics, as follows:

Russia

The number of books reported for Russia may have been high because of the number of languages in which the titles are published. While their figures were roughly comparable to those of the United States in raw terms, and in the 1980's overtook the United States, it is probably not accurate to use this measure as an indicator that the country is decidedly ahead of the United States in book production.

United States

There were no Unesco statistics for the United States. But in reviewing the data published in *Publishers Weekly,* he provides the subject breakdown shown on the next page.

Subject	1970	1975	1980	1982
Medicine	1,476	2,282	3,292	3,229
Science	2,358	2,942	3,109	3,124
Sociology, economics	5,912	6,590	7,152	7,449
Technology	1,141	1,720	2,337	2,328

Certainly United States production is increasing, as one would expect.

United Kingdom

There were no Unesco statistics for the United Kingdom either, and his own data for the country are not comparable to those of the United States. Instead of giving yearly totals in five-year increments for the same period, the averages are given for 1978/80 and 1981/83, with yearly totals for the 1980-83 years. So, we have the following, as provided by the Publisher's Association:

Subject	Ave. 1978-80	Ave. 1981-83
Aeronautics	183	227
Astronomy	138	149
Chemistry and Physics	736	711
Engineering	1,397	1,621
Geology & Meteorology	293	369
Medical Science	2,713	3,092
Natural Sciences	1,194	1,306
Political Sci. and Econ.	3,529	4,068
Psychology	726	755
Science, general	87	63
Sociology	1,050	1,122

While other categories were presented, these are the ones most closely allied with those of the present study. For the most part, U.K. publications did not decline either.

West Germany

Curwen cites West Germany as the "world's third largest publishing nation." Its output is given as follows:

Subject	**1977**	**1978**	**1979**	**1980**	**1981**	**1982**
Economics, Statistics, Social Science	4,043	4,456	5,312	6,658	5,598	5,326
Politics, Defence	1,224	1,204	1,406	1,737	1,256	1,238
Medicine	1,923	2,263	2,301	2,757	2,699	2,758
Natural Sciences	1,798	1,979	2,104	2,281	1,779	1,773
Mathematics	680	702	1,170	960	847	890

Like the world in general, West Germany peaked in production in 1980. Notice that these statistics are provided by the German book industry itself, and show a pattern different from the Unesco statistics, which admittedly are by language (German), not country. The Unesco statistics show a drop from 1977-78, and a jump from 1978-79. The German statistics show a jump from 1977-78, and substantial one from 1978-79. Unesco shows a decline from 1979-81, whereas the German statistics mirror the decline from 1980-81. They contradict the 1979-80 decline with an increase. Both sets of numbers do not restrict themselves to a given category of materials: they are both general numbers.

France

Curwen compared statistics from the Syndicat National des Éditeurs (SNE) with the Unesco statistics favorably, except for the social science area, wherein the Unesco data show an increase of several hundred percent for 1980-81, and the SNE reports a decline. Again, it is difficult to equate France with all French publishing.

The figures for subject relevant to this study are:

Subject	1970	1978	1979	1980	1981	1982
Scientific, technical	2,134	1,831	1,823	1,847	1,791	1,987
Social sciences	2,943	3,440	3,550	3,601	3,380	3,572

These also reflect the general worldwide pattern of an increase up to 1980, and decline and attempted recovery thereafter. The overall total for the country also shows this pattern. This is different from the secondary service patterns for French, which show an overall pattern of a slope down. Indeed, the secondary services include materials from other French-speaking countries such as Canada and Francophone African countries. Nonetheless, that should alone pull the numbers down.

Spain

Spain's pattern is a little different from the rest, in that it is generally flat between 1973 and 1979, and increasing thereafter. The data for relevant categories are:

Subject	1979	1980	1981	1982
Mathematics	570	541	598	700
Natural Science	1,056	1,321	1,164	1,253
Medicine, Public Hygiene	893	1,069	1,298	1,296
Engineering, Tech.	705	1,056	872	976

The data are from the Spanish *Anuario Estadistica*. Curwen suggested that the large number of medical titles might have been produced to supply former colonies. He notes that 5,000 of the 24,569 titles published in 1979 were translations, which is a substantial non-original amount. We, however, would have to know the number of translations for each year to be able to assess the figures more accurately. And translations are beyond the scope of this project.

Italy

The number of books (new titles, new editions, and reprints) published in Italy, according to the data shown in the *Annuario Statistico Italiano:*

Year:	**1978**	**1979**	**1980**
Titles	17,618	17,838	19,694

Curwen suggested in his analysis that subject categories are available because he notes trends in books in history, biography and the like, but he gives neither specific figures nor any mention of scientific works.

Japan

Data from the Research Institution for Publications of the Prime Minister's Office in Japan again showed increasing production, including production of first editions, which hover around 70-73% of all titles.

Year	**1977**	**1978**	**1979**	**1980**	**1981**
Titles	36,815	37,710	39,027	39,462	40,439

Curwen remarked that "Science and engineering are not as prominent as we might expect" but he does not give precise figures. He did indicate that social science, literature, art and industry account for 58% of the total.

China (People's Republic)

China, according to the Chinese State Council State Statistical Bureau, had been growing at rates ranging from 7.9-21.4%, the latter between 1980 and 1981. The specific figures were:

Year	**1977**	**1978**	**1979**	**1980**	**1981**
Titles	12,886	14,487	17,212	21,621	25,601

Curwen reported that the industry publishes in 14 "minority languages," but these represent 5% of all copies produced.

The secondary services are all increasing their coverage of Chinese materials, and the data suggest that there are more available as well.

India

As one of the largest publishers in English, India is of interest. According to the National Library in Calcutta, its output has been declining:

Year	**1976**	**1977**	**1978**	**1979**	**1980**	**1981**
Titles	21,922	19,629	18,584	16,392	17,168	11,562

International Serials Data Program Data

The International Serials Data Program (ISDS) is a "worldwide system for the identification and processing of periodicals and other serials publications" (ISDS, 1987). ISDS is structured as a series of regional and national centers, coordinated by an International Centre in Paris. Through it, International Standard Serials Numbers (ISSNs) are assigned to serials. These can be requested by publishers or managers of collections of serials. Bibliographic records are created when ISSNs are assigned, and these are compiled into the ISDS register. The database contained 351,772 records as of 1 January 1987, and grows at a rate of 40,000 titles per year. In terms of assignment of numbers, current and new titles are given priority, but "local interest" titles are included. There are 46 national centers, and two regional centers for the USSR and South East Asia.

ISDS records have not been analyzed yearly, according to staff at the National Serials Data Program in the Library of Congress. Data published in the *ISDS Newsletter,* however, indicate that in 1985, the ISDS database drew from 172 countries. In 1987 there were 183 represented. While the number of records in 1985 is presented in groups (those languages with over 20,000 records for example), it is possible to determine some increases by identifying countries that moved from one group to a higher one. We can see that Japan, for example, went from 8,000 to 11,500 records. Denmark, Finland, Hungary, Sweden, Belgium, Brazil, the Netherlands, Poland, Czechoslovakia, the German Democratic Republic, Indonesia, Ireland, Israel, Mexico, the Philippines and Portugal all increased. And, the number of languages represented grew from 113 to 125.

Summary

Given all the caveats about comparing different types of materials, and the inherent problems in trying to analyze data that are inconsistently and irregularly reported, the chapter described trends in production of various types of literatures from other sources. While these data are only suggestive of an increase in production of serials in non-English languages, the data do not suggest a decline.

CHAPTER 7

SUMMARY AND CONCLUSIONS

The Purpose of the Study

The purpose of this study was to explore the language distribution of materials included in selected on-line bibliographic databases between 1970 and 84 and to compare those data with those from other resources. The results describe the degree to which English has increased, decreased, or remained stable in relation to other languages, as reflected in the availability of bibliographic records in these databases.

Summary of the Data

Database Sources

Biosis. Chinese and Korean showed the largest increases in the last three years. A problem in analyzing the data was caused by the lack of assignment of language codes in records until the late 1970's. Nonetheless, each year more and more records were added to the database (despite a problem with 1983). Bulgarian, French, German, Italian, Japanese, Korean, and Polish followed the pattern of the database and could be interpreted as increasing their presence in the file; Russian and Chinese also increased but without declining in 1983. Portuguese and Spanish decreased in the later 1980's. However, because of the domination of English, all languages save Japanese and English decreased their proportion of the database.

Chemical Abstracts. Over the last three years, Chinese and Japanese showed the most dramatic increases. Over the last eighteen years, an increasing number of records were added to the database, and English increased its dominance of the file. Hungarian, Polish, Portuguese, and Romanian showed very erratic patterns and did not show a definite trend. Records for documents in English, Chinese, Czech, German, Spanish, and Japanese showed patterns of increase. Those for Bulgarian, French, Italian, Russian, and Ukrainian showed patterns of decline. As a proportion of the database, just English, Chinese, Czech, and Japanese showed an increase. The remainder of languages declined.

GeoRef. The database increased the number of records added to the file until 1982, when fewer and fewer records were added to the file each year. Records for documents in English, Chinese and Japanese increased in the database. Those for materials in French, German, Russian, and Spanish decreased, especially after 1982.

Country of publication data was also available for this file. The database contains materials from a wide variety of countries. Over the last three years, those from China and the Ukrainian SSR increased most dramatically. In raw numbers, only documents from China showed an increasing pattern. Canadian materials increased roughly until the middle 1980's, those from the United Kingdom did so until the early 1980's. Those from the Federal Republic of Germany and the Netherlands were too erratic to classify. Those from Japan were relatively stable after 1979. As a percent of coded records, however, materials from only the United States, Canada, China, the Netherlands, and the United Kingdom show patterns of increase. Overall, the number of non-English records has declined for this database. English records and materials from the United States dominated the file.

MEDLINE. An increasing number of documents were added to the database each year. Records for documents in Czech, French, German, Polish, Romanian, and Swedish declined with pronounced patterns. Records for documents in English, Chinese,

Danish, Dutch, Japanese and Spanish increased in numbers. Portuguese and Russian neither increased nor decreased their numbers. The pattern for records in Italian was simply erratic. As a proportion of the file, the only languages increasing were English, Japanese and Chinese.

Medline also includes country of publication data. Documents from Russia, Switzerland, and West Germany decreased. The figures for those from France were largely stable. Those from England, Italy, Japan, the Netherlands, and the United States increased. As a proportion of the file, those from the United States increased the most dramatically.

NCJRS. The general pattern for NCJRS is a mountain-shape, with general growth until 1978 and subsequent decline. English increasingly is the major language for the file, and Dutch, French, German, Italian, Japanese, Polish, Spanish, and Swedish mirror the pattern for the file. They cannot be seen as increasing in proportion. Even as a percentage of the file, all languages except English are not increasing either.

Country of publication statistics show materials from Australia, Belgium, Canada, France, India, Italy, Japan, the Netherlands, Poland, Sweden, Switzerland, the United Kingdom, the United States, and West Germany. None were shown to be increasing, and the United States contributed the bulk of the file consistently.

The condition of the NCJRS file directly relates to the mandates from the Department of Justice, which instructed the producers of the file to reduce the number of records entered. The producers in turn restricted the number of non-English materials to save costs. Because of their difficulty of acquisition and handling, they were the simplest to cut. The producers were anxious to reverse this trend, and the figures for the last two years may indicate success.

Oceanic Abstracts. Over the last three years, the proportion of non-English materials decreased dramatically. In the original study, twelve languages were needed to account for 99% of the database. The updated data showed that only English, French, German, Japanese, Spanish, and Russian were needed. Even though they were small in number, increases in Japanese and Chinese were dramatic. The database did not add an increasing number of records each year: even this pattern was erratic. There were no definite trends for Italian, Japanese, Korean, Norwegian, Polish, Portuguese, and Spanish. Records for materials in Chinese and Russian could be interpreted as increasing, but 1987 was not a good year for the latter.

Oceanic Abstracts also include summary languages, and these increased, but only because the second half of the period 1973-84 is stronger than the first. The strongest languages for the summaries were English, French, German, Italian, Japanese, Spanish, and Russian. There were insufficient records to trace which languages were increasing or decreasing as a summary language for a specific language. Clearly, however, English and the language of the document itself were the languages of choice for summaries.

PAIS. The database added a decreasing number of records to the file each year. And, the few non-English languages included (French, German, Italian, Spanish, and Portuguese) all decreased in number. These patterns were echoed in their proportion of the file, showing no increases. Spanish showed a slight increase. Only English and Spanish increased their representation in the database.

PsycInfo. Over the last three years, Spanish and Chinese showed the most dramatic increases. However, only eight languages in 1989 rather than the earlier twelve were needed to account for 99% of the entire database. Except for 1979, 1987 and 1988, an increasing number of records were added to the database each year. The patterns over the last eighteen years could be described as increasing for Dutch, Italian, Portuguese and Spanish. Patterns for French, German, and Japanese followed the database pattern, and can be described as increasing. Those for Czech, Polish, and Serbo-Croatian were erratic. As a proportion of the database, English again dominates the file. No language except English increased its proportion.

The database producer could not account for any specific trends in the inclusion of languages, but its staff contended that they do attempt to cover the worldwide literature.

Aggregate Data

The following chart shows the total non-English records in these databases, year by year. Despite the decreases in individual databases, their representation increased significantly by 50% over the period of study.

Non-English Records in Eight Databases:
Records by Year

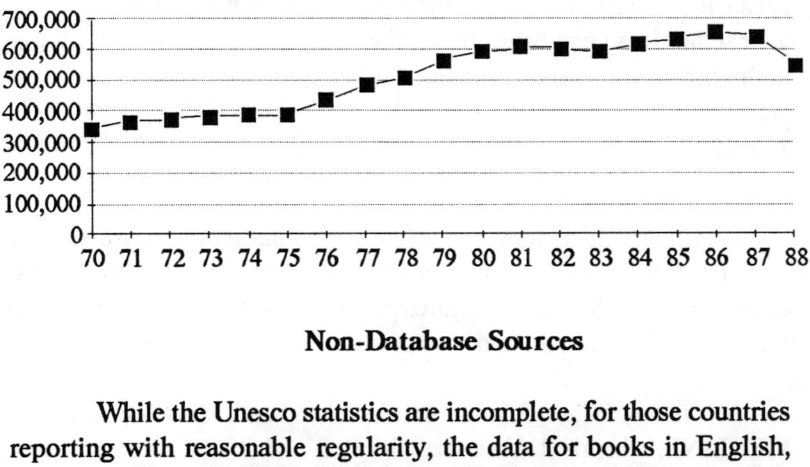

Non-Database Sources

While the Unesco statistics are incomplete, for those countries reporting with reasonable regularity, the data for books in English, French, German, Russian and Spanish suggest that publishing in these languages has not been declining between 1970 and 1984.

The Curwen analysis of publishing statistics from national sources such as publishing associations and national accounts showed that while publications from the United States and the United Kingdom declined in general science for the period, they showed upward trends for specific sciences. Statistics for West Germany, Spain, Italy, Japan and China in general have increased. For France, social science statistics have increased, and science and technology books were slightly down but recovering as of 1981 to 1982. Statistics for India have declined.

While the data from the International Serials Data System are scant, they indicate increased coverage of serials in more countries, and in more varied languages.

The Unesco and Database Statistics Compared

For Russian, German, Spanish, French, and English, it is possible to compare database-reported coverage with trends in the book literature as reported by Unesco participating countries. The databases are presented in two parts (larger and smaller databases) because the relative sizes of the files prevents them from showing their trends in a composite chart. The lines for the files tend to clump at the bottom.

French

Unesco data show that while the actual numbers are erratic, even the low estimates do not show a decrease in production. The chart showing the actual figures reported, with the estimates for possible high, medium, and low production, is on the next page.

The database figures, on the other hand, show declines for MEDLINE, Chemical Abstracts, GeoRef, and PAIS. Only Oceanic Abstracts and PsycInfo could be considered increasing, and then only slightly.

The charts for French begin on the next page.

Unesco Statistics for French Documents:
Actual Statistics with Estimates

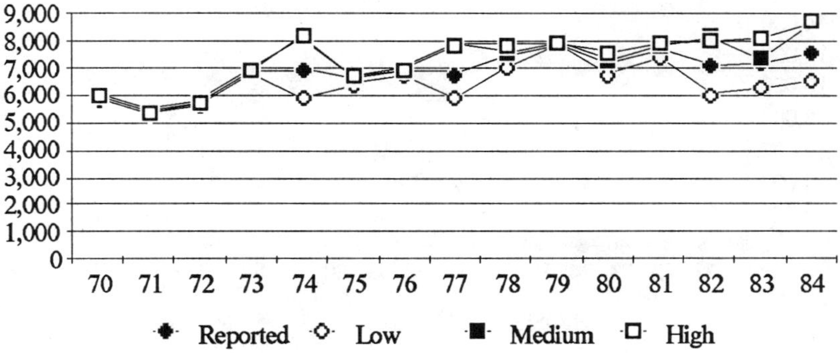

French Records in Smaller Databases, by Year

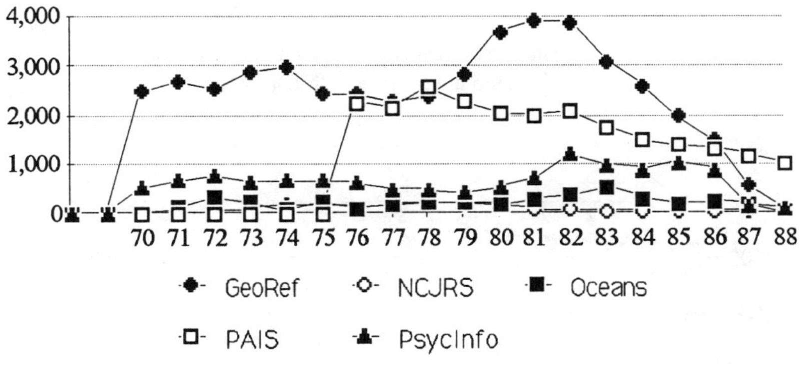

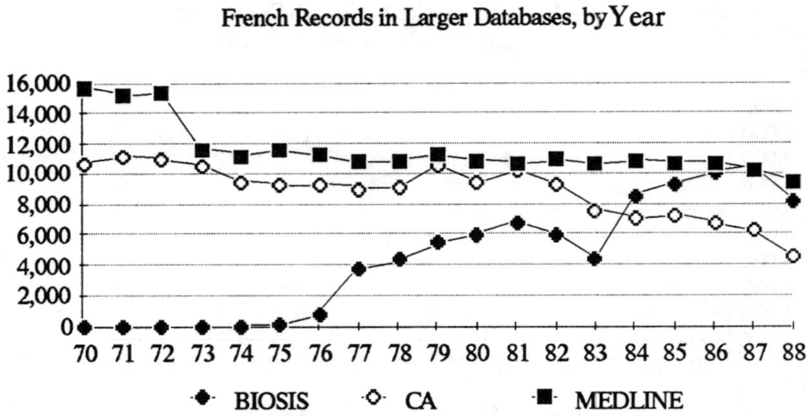

German

The Unesco data show a consistent, steady increase, with the exception of 1983.

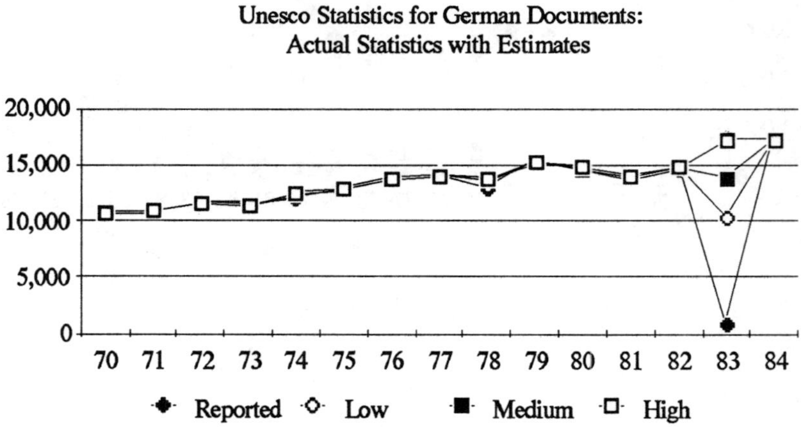

German Records in Smaller Databases, by Year

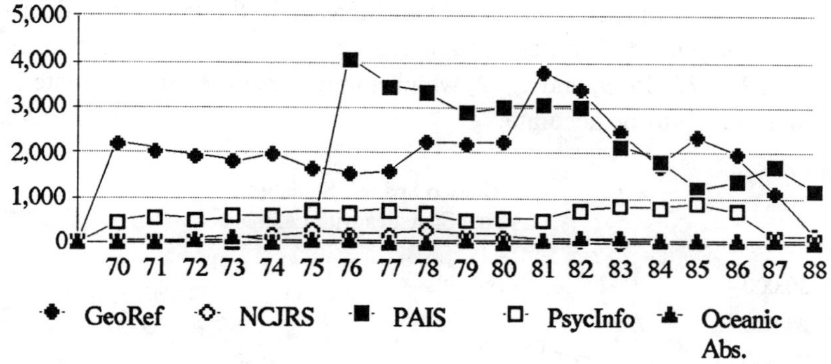

For the databases BIOSIS, Chemical Abstracts, and PsycInfo exhibit patterns that can be interpreted as increasing. MEDLINE, GeoRef, NCJRS, and PAIS show declining patterns.

German Records in Larger Databases, by Yeaır

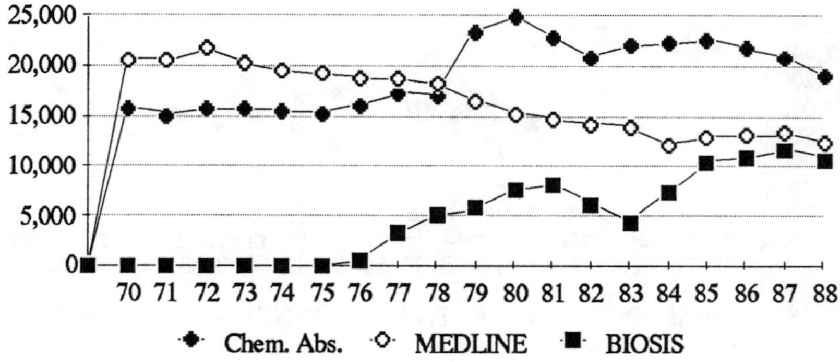

Spanish

The Unesco data suggest an increase in the number of documents produced. The actual statistics are equivalent to high estimates for 1970-77, 1979, and 1982, which suggest that the high estimate is most likely to be accurate.

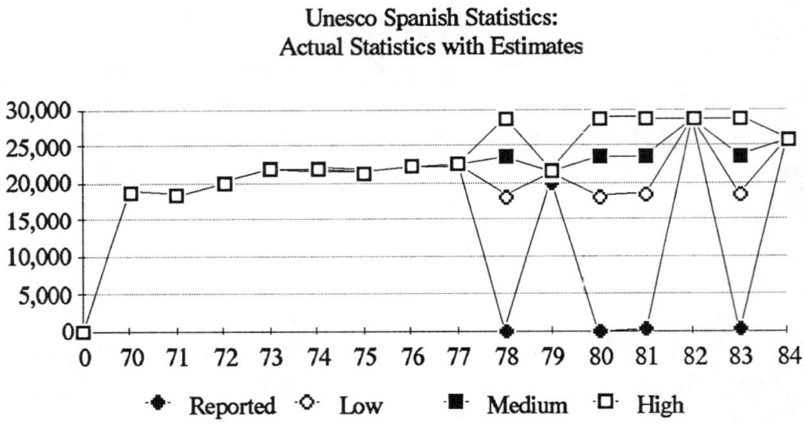

On the other hand, the BIOSIS, Chemical Abstracts, PsycInfo, and Oceanic Abstracts show recent declines.

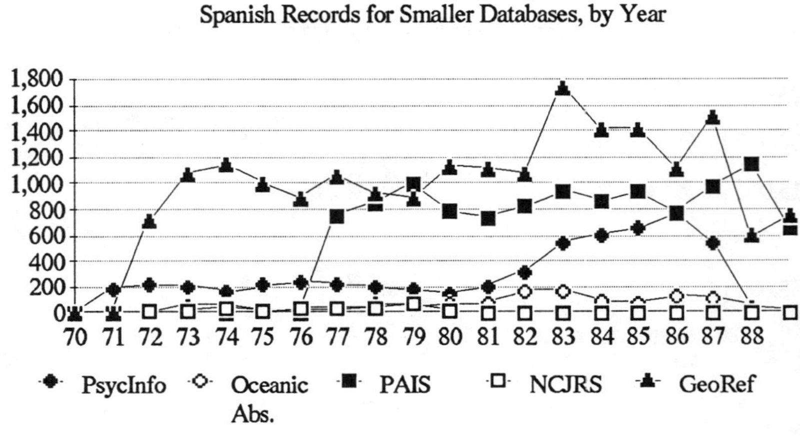

Spanish Records in Larger Databases, by Year

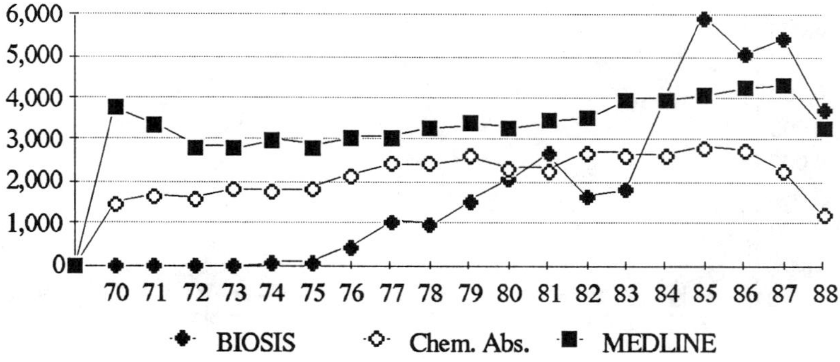

Russian

The Unesco data are remarkably consistent for countries reporting production of documents in that language. The pattern is simply flat and stable.

Unesco Russian Statistics:
Actual Statistics with Estimates

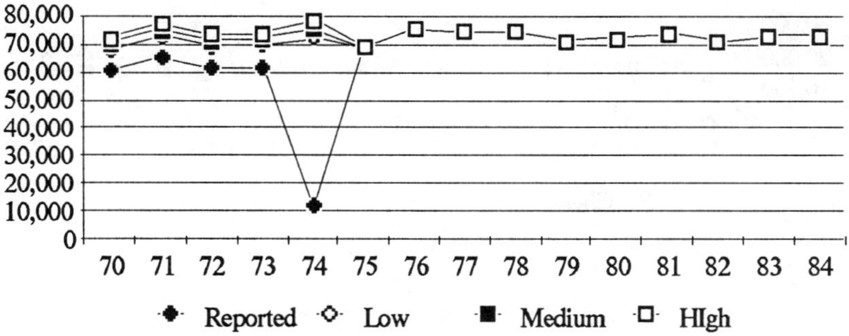

For the databases, PsycInfo is highly erratic but shows no overall increase in coverage. BIOSIS shows a slight increase, but that most probably is a reflection of coding practices. MEDLINE generally follows the Unesco pattern of stability. Oceanic Abstracts shows some increase in coverage.

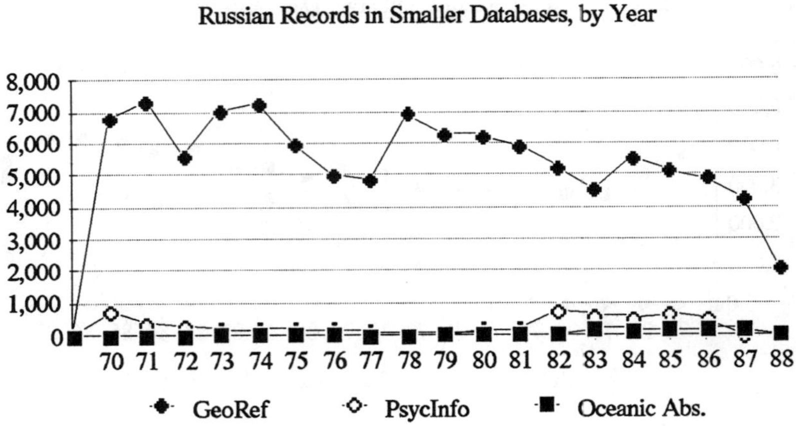

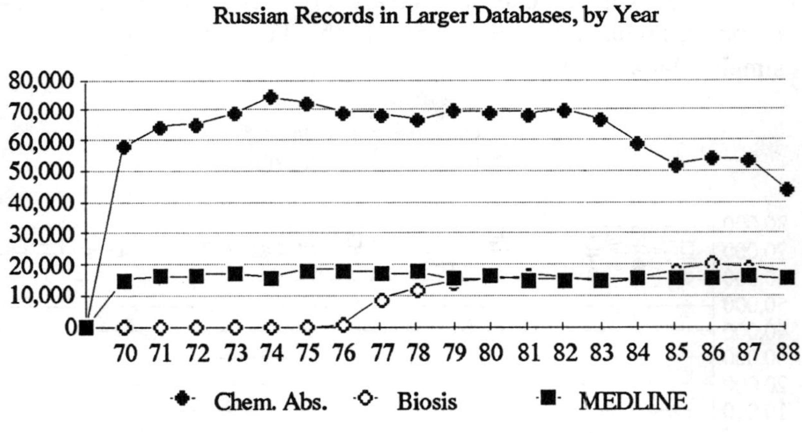

English

The Unesco reported statistics show a decline in production in English, the estimates suggest a more stable pattern. Note that figures for the United States, United Kingdom, and other English-speaking countries are absent.

Unesco Statistics for English Documents:
Actual Statistics with Estimates, by Year

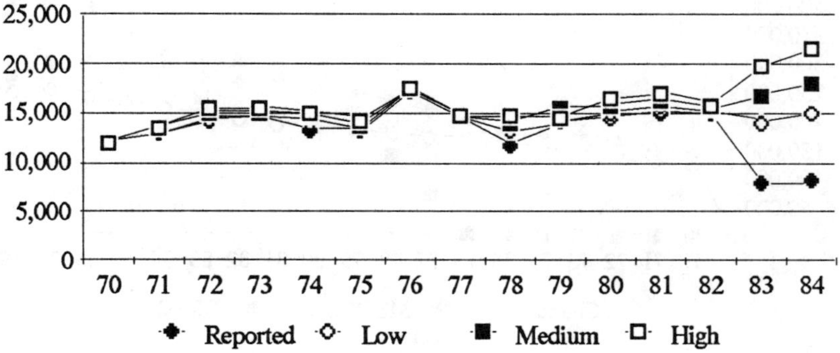

The NCJRS database shows its usual mountain pattern, PAIS and BIOSIS show large increases which may be coding difficulties. The GeoRef, PsycInfo, Chemical Abstracts, and MEDLINE show a rate of increase higher than the Unesco patterns.

English Records in Smaller Databases, by Year

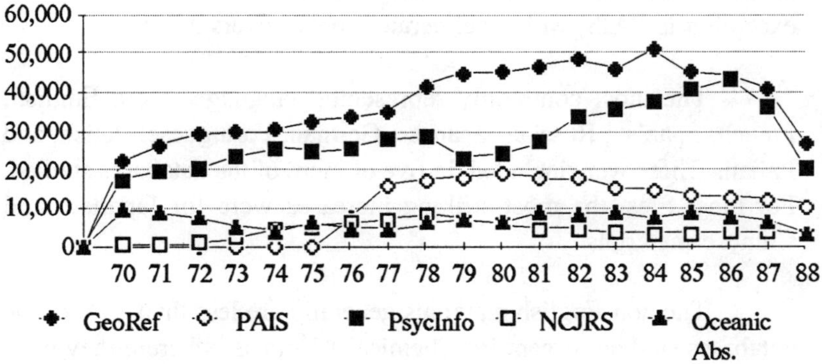

English Records in Large Databases, by Year

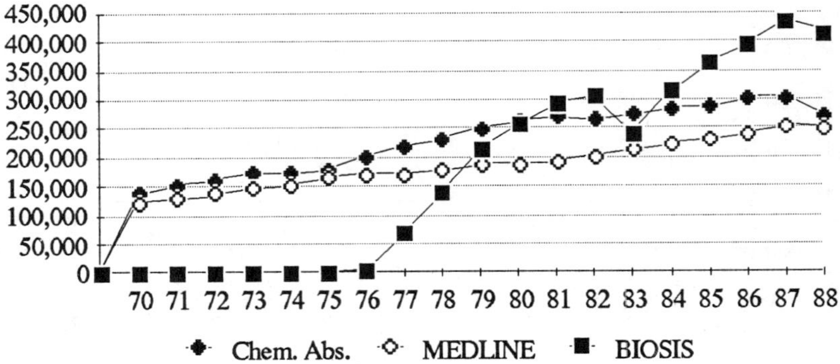

Research Questions

With all of the data presented, it is now possible to answer some basic questions about language trends in the databases studied.

1. What languages are represented in the different databases, and what proportion of each database do they represent?

This information is presented in Chapter 6 and in the charts in the Appendices. While many of the databases include documents in a wide variety of languages, some over a hundred different languages, in most cases only a handful comprise 99% of the database. The exception is PAIS, which deliberately only covers a few.

The most commonly represented languages were English, French, Spanish, Russian Japanese, German, Portuguese, Polish, and Italian. These were included in five or more of the databases studied. For most files, the most striking increases were for Japanese and Chinese materials.

The non-English materials generally are less than 15% of the databases studied, except for Chemical Abstracts, wherein they were approximately 35%. This is a significant change from the early 1970's when non-English materials commonly were 40% to a half of Chemical Abstracts, MEDLINE, GeoRef, and PAIS.

2. What countries of publication are represented in the different databases, and what proportion of each database do they represent?

Only three databases include country information (GeoRef, MEDLINE, and NCJRS). These are detailed in Chapter 6. Overwhelmingly the United States is the most heavily represented, and the trends have shown that the most concentrated increases have been for Japanese and Chinese materials.

3. What is the proportion of abstracts and summaries in languages other than the text of the document?

Only Oceanic Abstracts included summary language information. It showed that by far English and the language of the document itself are the languages of preference for summaries.

4. How have the first two (languages and countries of publication) varied between 1970 and 1984? What languages have increased, decreased, or remained stable?

This is detailed in Chapter 6.

The following final table, Table 7-1 shows the languages studied for each database, and a plus (+) if the language has increased, a minus (-) if its representation has declined, and an E if the language has been either too erratic to classify or has been stable. These assessments are based on the raw numbers, not percentages in the file. The order was determined by assigning a plus a 2, a minus a -2, and an E a 0.

Table 7-1: Trends in Languages in Databases

	Database							
Lang.	**Bio.**	**CA**	**Geo**	**Med**	**NCJ**	**OCE**	**PAIS**	**Psyc**
Chinese	+	+	+	+		+		
Japanese	+	+	+	+	-	E		+
Bulgarian	+	-		+				+
English	+	+	+	+	-	-	-	+
Spanish	-	+	-	+	-	E	+	+
Korean	+					E		
Slovak		+		E				
Dutch				+	-			+
Danish				+				
German	+	+	-	-	-	E	-	+
Norwegian				-				
Romanian		E	-					
Portuguese	-	E		E		E	-	+
Serbo-Croatian					-			E
Ukrainian		-						
Czech		+	-	-				E
Hungarian		E		-				-
Russian	+	-	-	E		+		E
Polish	+	E		-	-	E		E
Italian	+	-		E	-	E	-	+
Hungarian		E		-				-
Swedish				-	-			
French	+	-	-	-	-	+	-	+

+ = increased, - = decreased, E = too erratic or stable to classify

This table roughly shows the languages that have increased and decreased over the period of study. English remains the pre-eminent language for these databases. Note, however that the traditional languages that are acknowledged as the major languages for science (French, German, Russian, and Italian) are all decreasing more often than they are increasing, if at all. The only other traditional, strong language is Spanish. Note also the consistent interest in Chinese and Japanese materials.

The same kind of chart for data on country of publication can be presented, for the databases that included it:

Table 7-2: Country of Publication Trends in Databases

Country	Database Bio.	CA	Geo	Med	NCJ	OCE	PAIS	Psyc
U.S.			+	+	-			
Netherlands			E	+	-			
UK			-	+	-			
China			+					
Japan			+					
Italy				+	-			
Australia					-			
Belgium					-			
Canada					-			
India					-			
Sweden					-			
Poland				-	-			
Switzerland				-	-			
France			E	E	-			
W. Germany			E	-	-			
Russia			-	-	-			

+ = increased, - = decreased, E = too erratic or stable to classify

While these statistics reflect the strength of publishing from the United States and United Kingdom, they do not reflect conventional figures for Germany, France, and Russia. These are admittedly weak data, however, since effectively they come from only two databases because the non-U.S. publications were deliberately decreased.

5. How do these statistics (the database statistics) compare with statistics for actual literature production?

The only non-database statistics available for review are the Unesco and Curwin figures, which concern books. These suggest that materials in French, German, Spanish, and Russian, and publications from France, Spain, Germany are increasing rather than decreasing.

Note that we do not have type of document statistics for all databases. We know that MEDLINE is 100% journal articles, and books are less than one percent of the file for BIOSIS, Chemical Abstracts, and PsycInfo. They can be estimated at 5% for Criminal databases. We know that MEDLINE is 100% journal articles, and books are less than one percent of the file for BIOSIS, Chemical Abstracts, and PsycInfo. They can be estimated at 5% for Criminal Justice, and 10% for GeoRef. They appear to be absent for Oceanic Abstracts and PAIS, but that is uncertain because of coding difficulties.

Therefore, in a sense, we are comparing two different types of materials, books and journals (and other materials). Comparing these data with journal title production will be discussed below, in further research.

6. Do languages vary by document type (e.g., journal article, monographic publication, conference proceeding)?

This cannot be answered because the data were gathered by searching rather than examination of the actual records. For the one file for which this was available (Oceanic Abstracts), there was just not enough activity to report when languages were compared with document types.

7. Do databases in the social and physical sciences share any trends?

They do not appear to do so. Examining the charts above indicate no regular pattern for language or country by broad discipline of database.

8. How have database provider policies and practices affected these trends?

These practices appear to be the primary set of factors governing the increase or decrease of documents in the databases of study. The staff of the PAIS file indicated that they are dependent upon the acquisitions of the New York Public Library for their foreign language materials, although they do have opportunity to make suggestions. The staff of PsycInfo clearly indicated that it is the availability measure, and responded by cutting non-English materials. The Oceanic Abstracts added contributions from input centers around the world, and indeed this showed up as an increase in the number of non-English publications in the file.

Conclusions

The number of databases over the last twenty years have increased dramatically. Eight of the major, international files were examined for language content, to assess the representation of English and other languages. Country of origin data were collected and reviewed where available. Overall, English and materials from the United States have increased in numbers and in proportion to other materials in these databases. Some languages such as Spanish, Chinese, and Japanese have made gains in these files. Significant languages traditionally thought of as scientific languages have suffered declines in coverage. These include Russian, German, and Italian.

From discussions with the database providers, it seems that the variations in the files are due primarily to database provider practices, or external events not under their control. These include the availability of indexers competent in the languages needed, or directives from the database sponsor. While only one instance was found, it seems also that coding problems in the records themselves can also affect the picture that a database provides.

The secondary services studied intend to cover the world's literature in their given areas. Chemical Abstracts staff were quite confident that they were successful. Criminal Justice staff knew quite well that they were not. PAIS staff indicated that there is more out there, but documents omitted would not be as relevant to the file as those currently covered.

Statistics on book production from a variety of sources suggest that the proposition that the actual numbers of materials in these languages may not be true. As noted above, this is largely comparing books and journals, and it could be said that scientific writings in journals are increasingly being published in English but books are retaining the language of the country of origin. It will take further research to form a more definitive answer.

At the very least, we can say that for these databases (other than Oceanic Abstracts and NCJRS), the proportion of materials in languages other than English has declined, but the actual number overall has increased. And, we can say that English certainly has become the language of the secondary services- -but not necessarily that of science.

Appendix 1

1.1 IRL Life Sciences Collection (Information Retrieval Limited)

These records were provided to Victor Rosenberg by Information Retrieval Limited as a part of his study in 1981.

Language	*1978*	*1979*	*1980*	*Total*
Russian	4,230	4,347	4,672	13,249
German	4,600	4,075	4,329	13,004
French	4,350	4,225	4,411	12,986
Italian	1,104	1,082	1,087	3,273
Spanish	952	847	776	2,575
Japanese	736	859	766	2,361
Czech	449	475	514	1,438
Portuguese	335	267	306	908
Danish	171	145	184	500
Chinese	113	77	138	328
Swedish	75	54	81	210
Bulgarian	64	43	91	198
Serbo-Croatian	53	29	51	133
Norwegian	47	39	26	112

1.2 RILM Records
(Répertoire International de Littérature Musicale)

These records were provided to Victor Rosenberg by the RILM from their DIALOG file as a part of his study in 1981.

Language	*1972*	*1973*	*1974*	*1975*	*1976*	*1977*	*Total*
German	1,085	1,346	1,128	1,610	6,422	890	12,481
French	256	236	287	273	1,140	187	2,379
Russian	298	247	204	245	609	50	1,653
Italian	177	118	88	147	423	122	1,075
Japanese	119	201	136	109	368	17	950
Czech	46	85	32	20	485	16	684
Spanish	9	17	41	53	184	7	311
Swedish	50	37	58	59	77	17	298
Portuguese	25	6	20	17	65	3	136
Danish	19	4	3	62	29	0	117
Norwegian	15	5	18	38	24	13	113
Serbo.	12	4	0	2	29	2	49
Bulgarian	0	0	0	0	41	0	41
Chinese	0	0	0	8	8	0	16
Arabic	0	1	0	1	0	0	2

1.3 Chemical Abstracts

These records were provided to Victor Rosenberg by the Chemical Asbtracts Service as a part of his study in 1981.

Chemical Abstracts Records for Arabic Documents, Records by Year (Rosenberg Data)

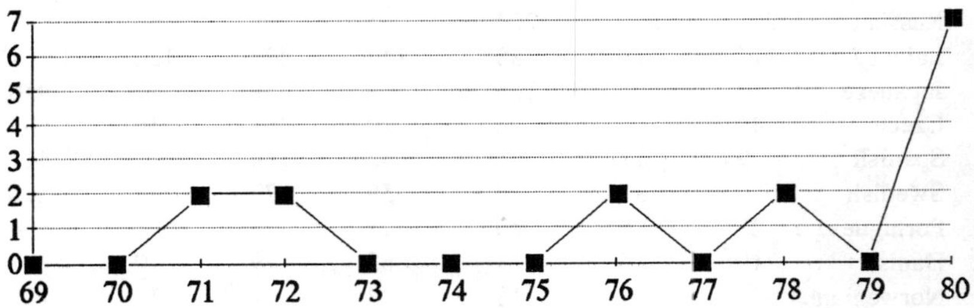

Chemical Abstracts Records for Bulgarian Documents, Records by Year (Rosenberg Data)

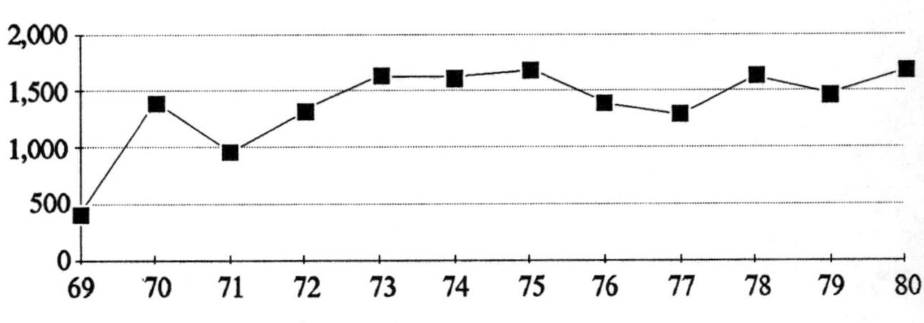

Chemical Abstracts Records for Chinese Documents,
Records by Year (Rosenberg Data)

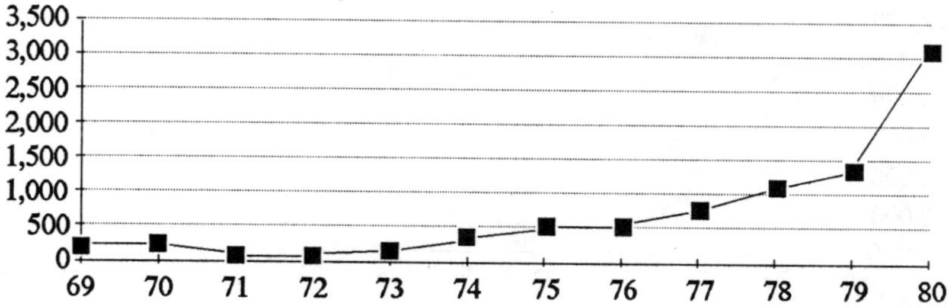

Chemical Abstracts Records for Czech Documents,
Records by Year (Rosenberg Data)

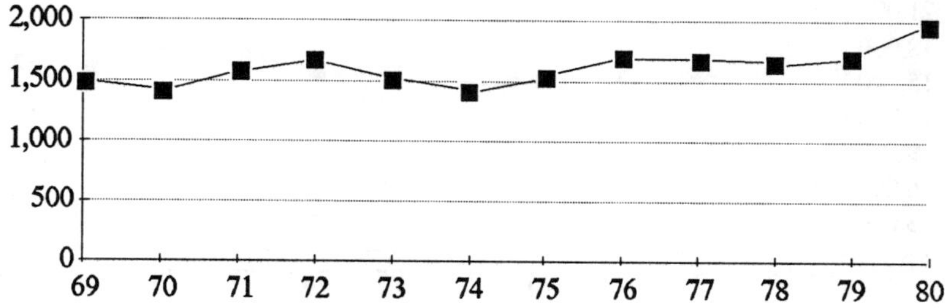

Chemical Abstracts Records for Danish Documents,
Records by Year (Rosenberg Data)

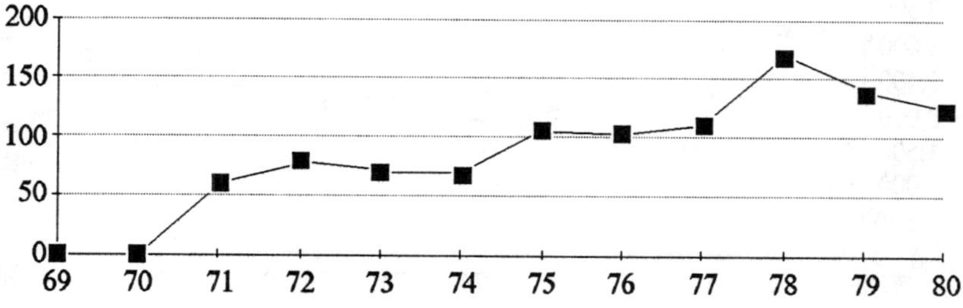

Chemical Abstracts Records for French Documents, Records by Year (Rosenberg Data)

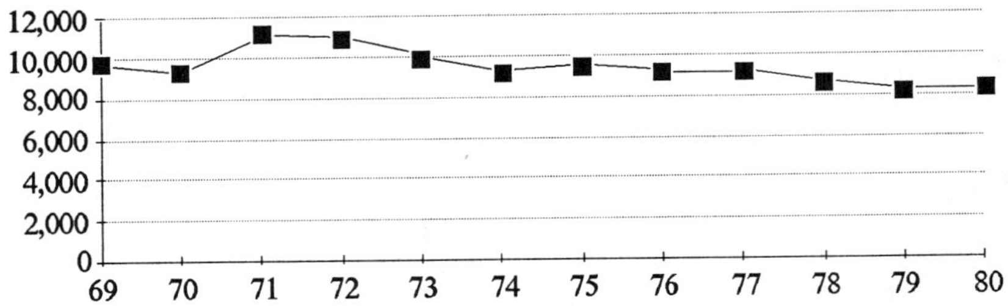

Chemical Abstracts Records for German Documents, Records by Year (Rosenberg Data)

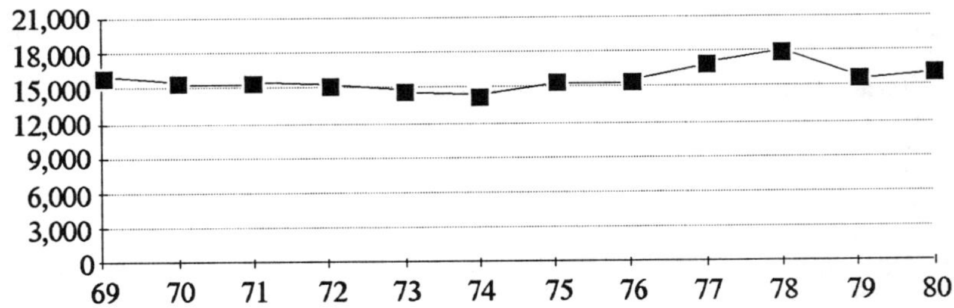

Chemical Abstracts Records for Italian Documents, Records by Year (Rosenberg Data)

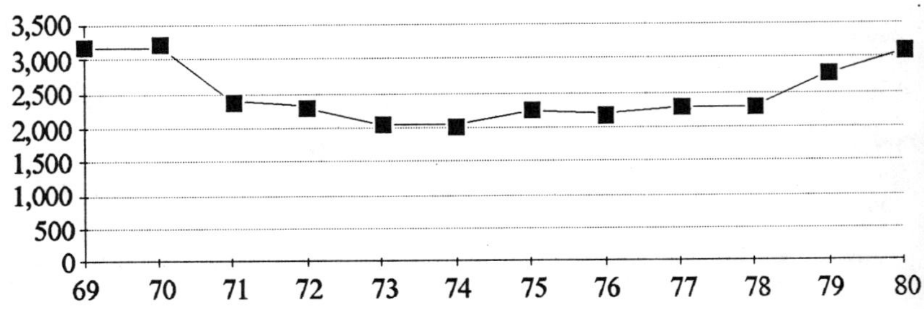

Chemical Abstracts Records for Japanese Documents, Records by Year (Rosenberg Data)

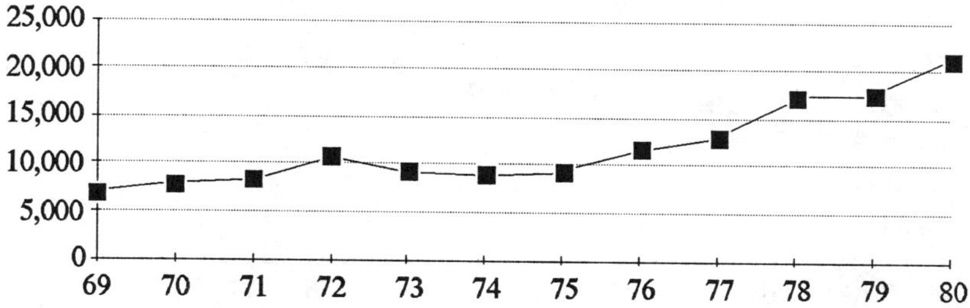

Chemical Abstracts Records for Norwegian Documents, Records by Year (Rosenberg Data)

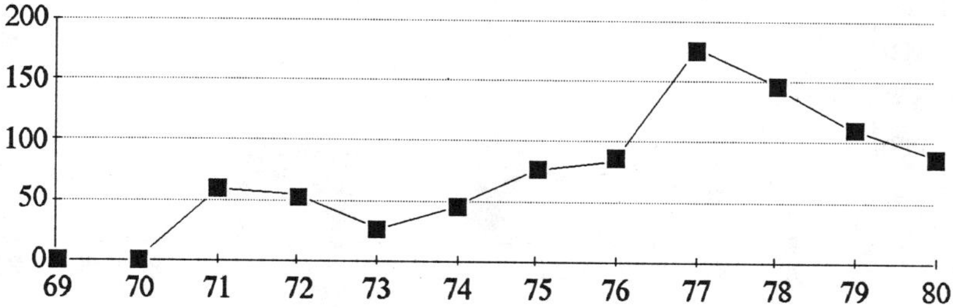

Chemical Abstracts Records for Portuguese Documents, Records by Year (Rosenberg Data)

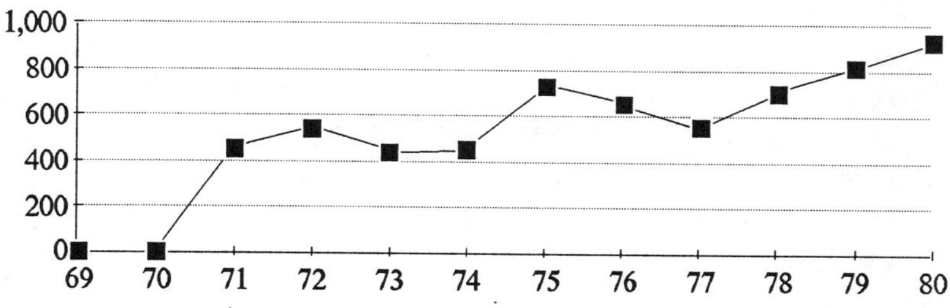

Chemical Abstracts Records for Russian Documents,
Records by Year (Rosenberg Data)

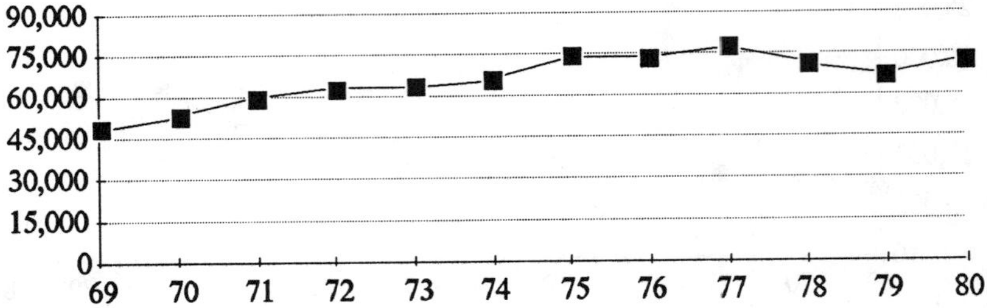

Chemical Abstracts Records for Serbo-Croatian Documents,
Records by Year (Rosenberg Data)

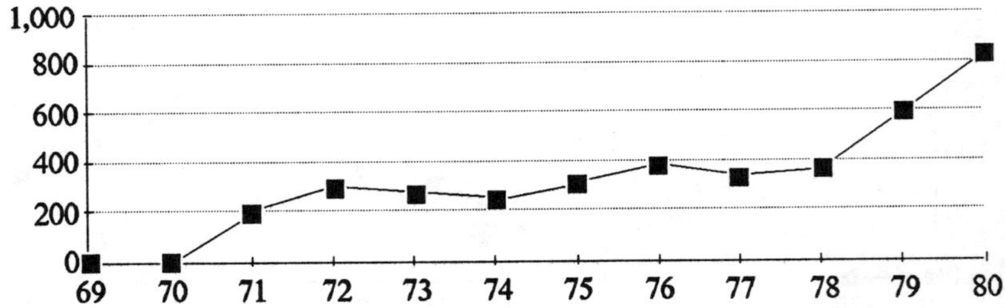

Chemical Abstracts Records for Spanish Documents,
Records by Year (Rosenberg Data)

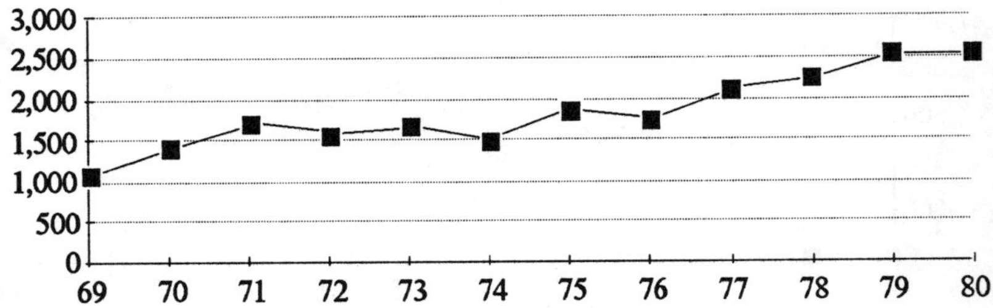

Chemical Abstracts Records for Swedish Documents,
Records by Year (Rosenberg Data)

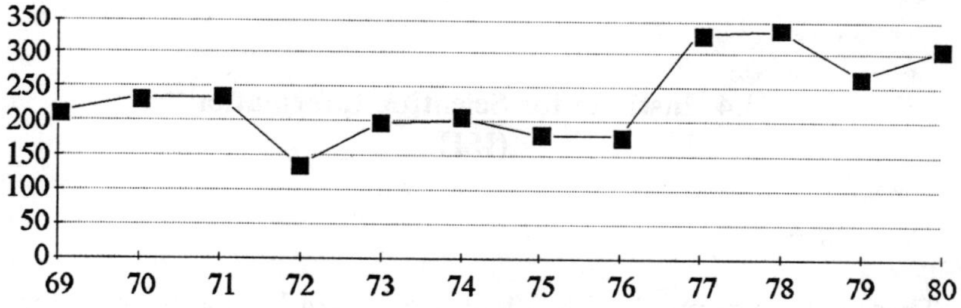

1.4 Institute for Scientific Information (ISI)

These data were provided to Victor Rosenberg by the Institute for Scientific Information as a part of his study in 1981.

ISI Records for Bulgarian Documents,
Records by Year (Rosenberg Data)

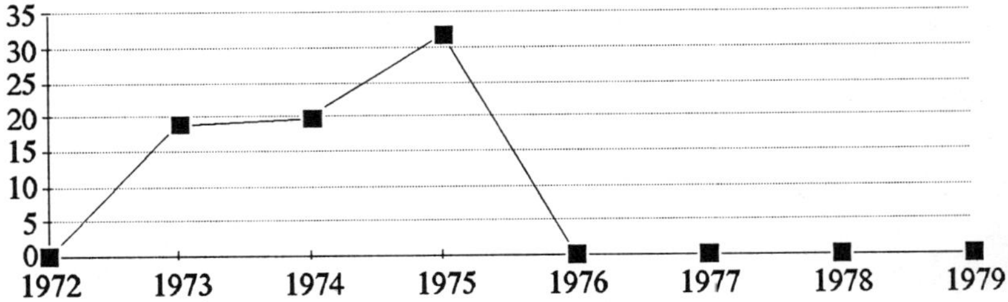

ISI Records for Chinese Documents,
Records by Year (Rosenberg Data)

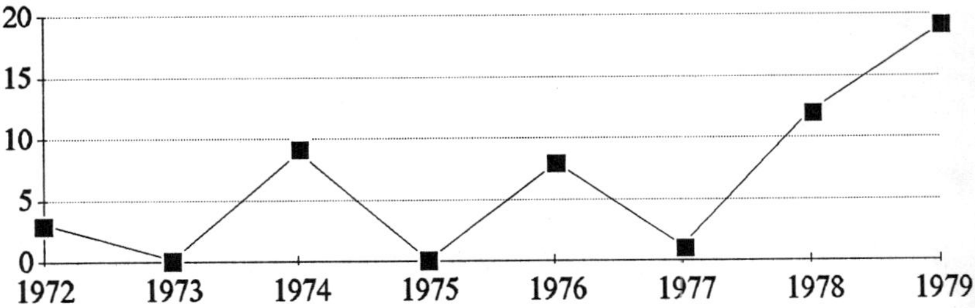

ISI Records for Czech Documents, Records by Year (Rosenberg Data

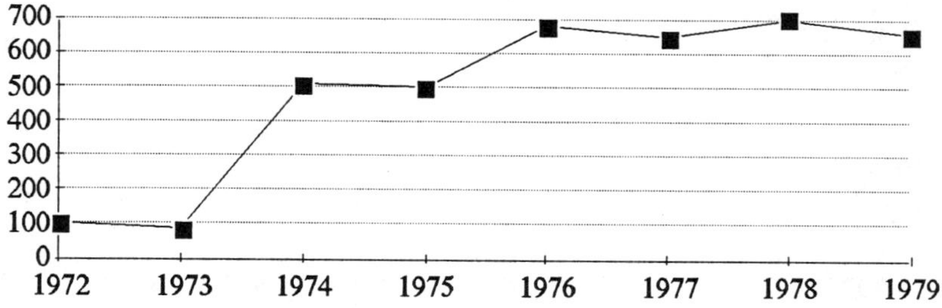

ISI Records for Danish Documents, Records by Year (Rosenberg Data)

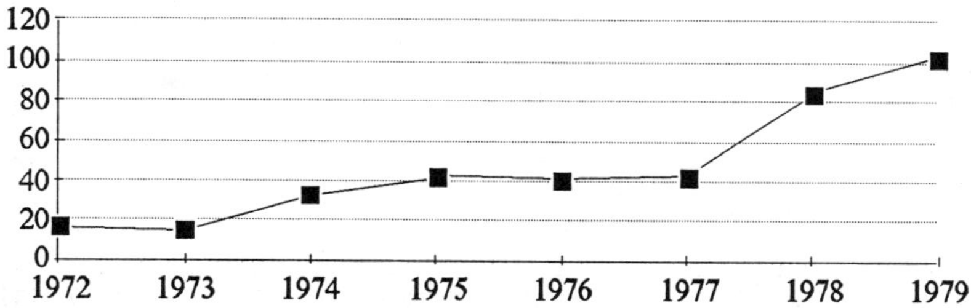

ISI Records for French Documents, Records by Year (Rosenberg Data)

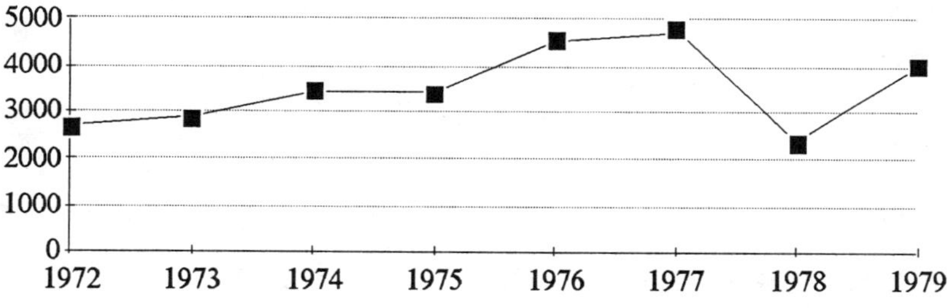

ISI Records for German Documents,
Records by Year (Rosenberg Data)

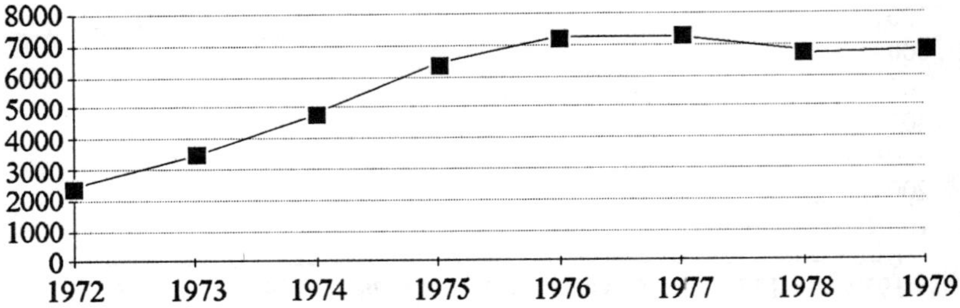

ISI Records for Italian Documents,
Records by Year (Rosenberg Data)

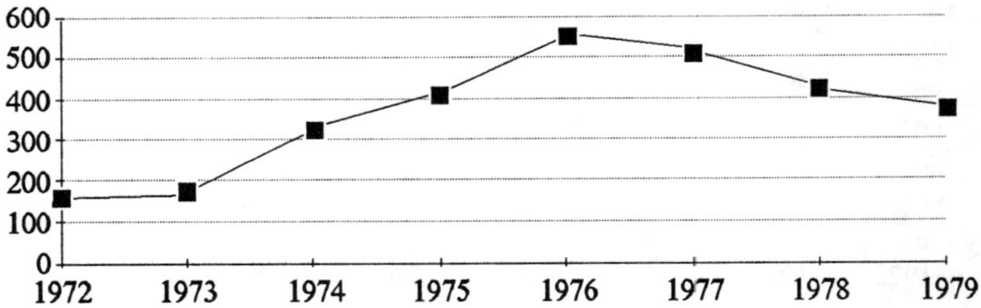

ISI Records for Japanese Documents,
Records by Year (Rosenberg Data)

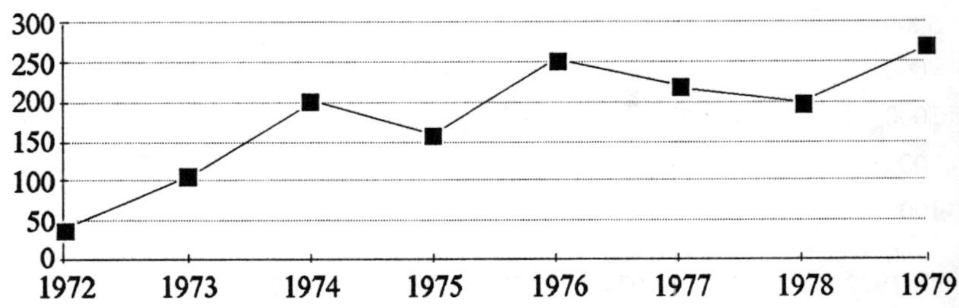

ISI Records for Norwegian Documents,
Records by Year (Rosenberg Data)

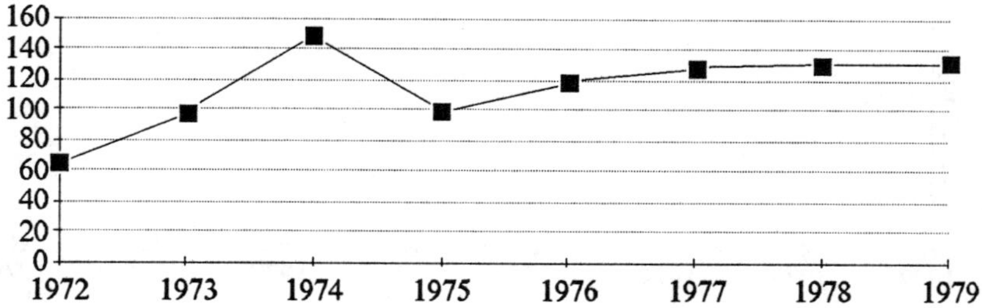

ISI Records for Portuguese Documents,
Records by Year (Rosenberg Data)

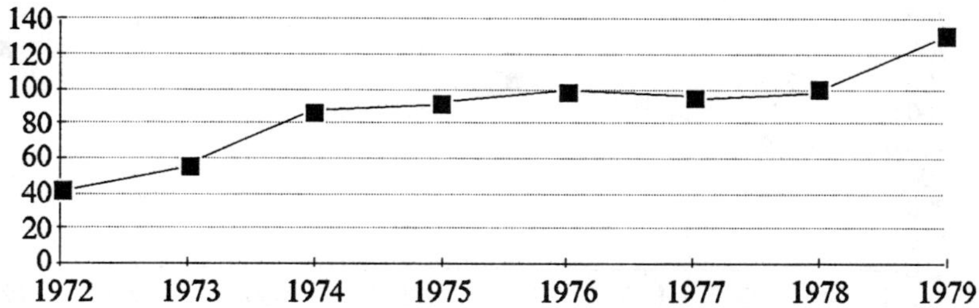

ISI Records for Russian Documents,
Records by Year (Rosenberg Data)

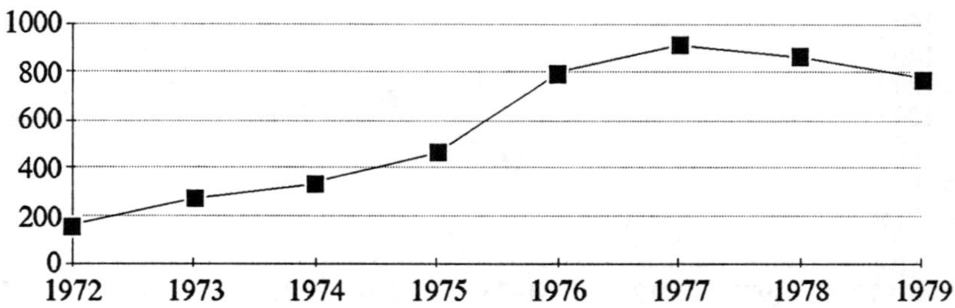

ISI Records for Serbo-Croatian Documents,
Records by Year (Rosenberg Data)

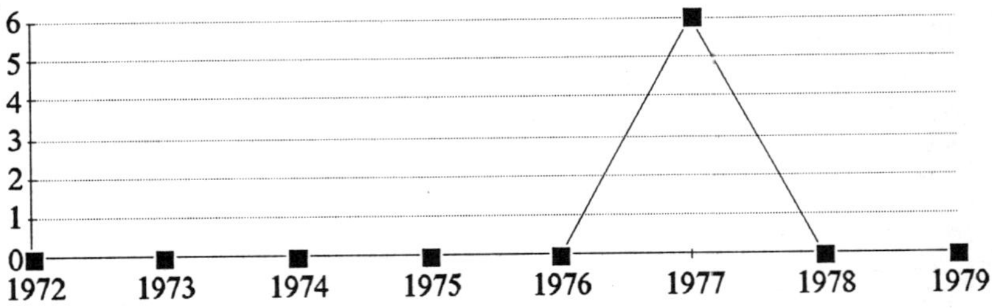

ISI Records for Spanish Documents,
Records by Year (Rosenberg Data)

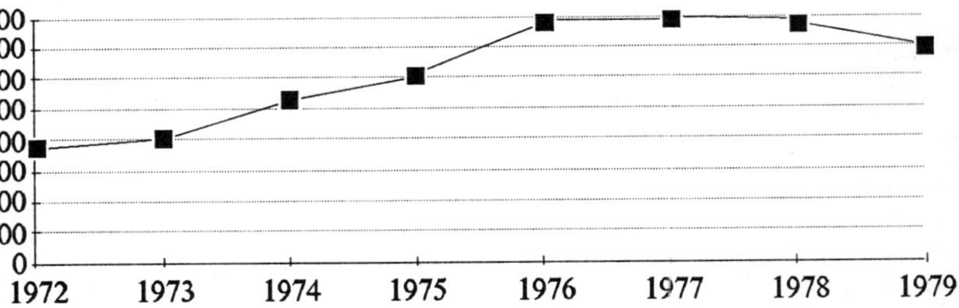

ISI Records for Swedish Documents,
Records by Year (Rosenberg Data)

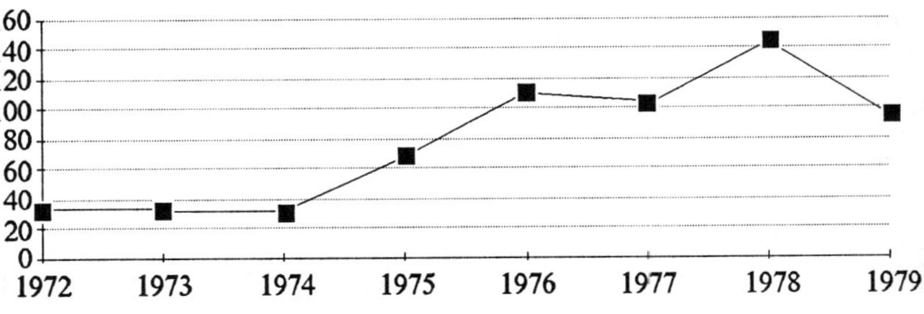

Appendix 2

Computer Horizons Data
Prepared for the
National Science Foundation, 1985

U.S. Percent of the World's S&T Literature:
Percent for All Fields, by Year
(No data for empty years)

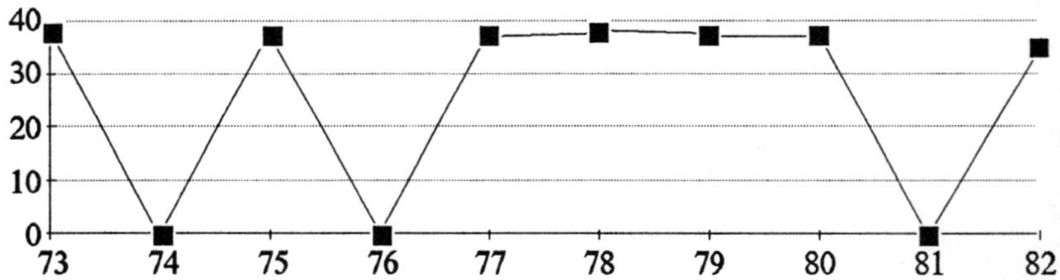

U.S. Percent of the World's S&T Literature:
Percent for Biology, by Year
(No data for empty years)

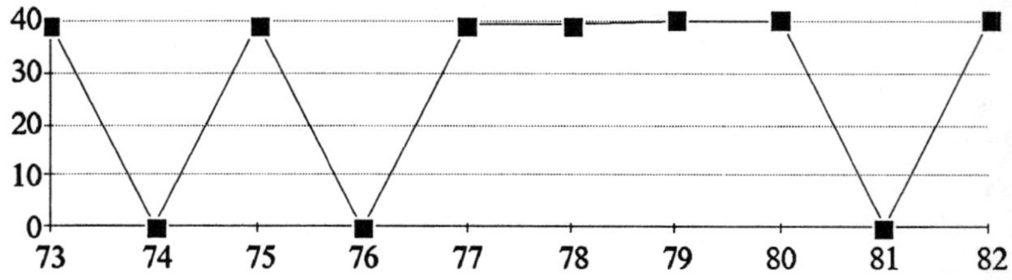

U.S. Percent of the World's S&T Literature:
Percent for Chemistry, by Year
(No data for empty years)

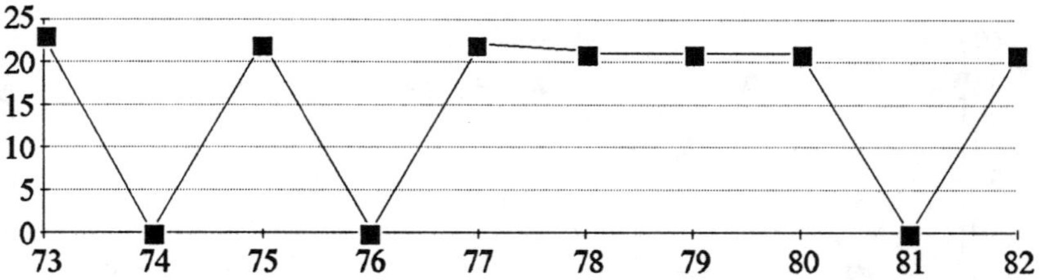

U.S. Percent of the World's S&T Literature:
Percent for Clinical Medicine, by Year
(No data for empty years)

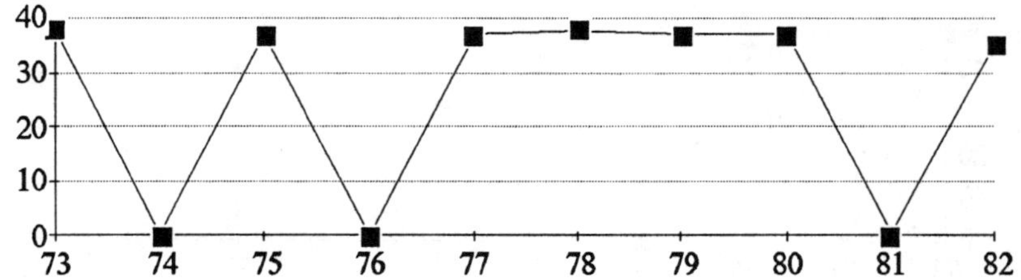

U.S. Percent of the World's S&T Literature:
Percent for Earth and Space Sciences, by Yeaı
(No data for empty years)

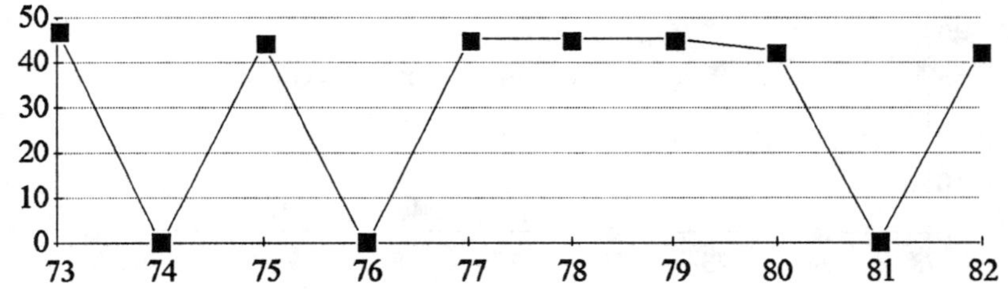

U.S. Percent of the World's S&T Literature:
Percent for Engineering and Technology, by Year
(No data for empty years)

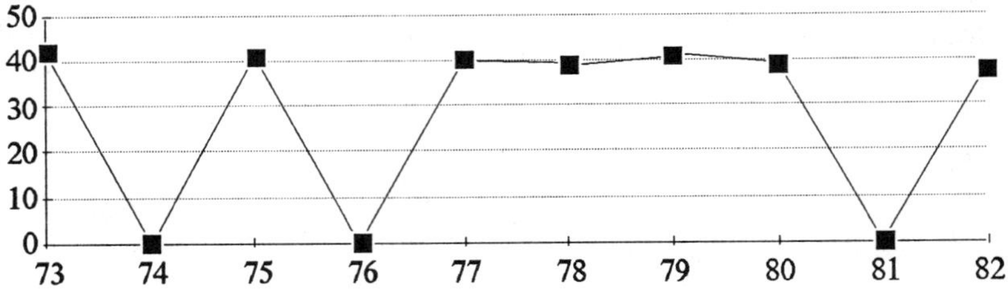

U.S. Percent of the World's S&T Literature:
Percent for Mathematics, by Year
(No data for empty years)

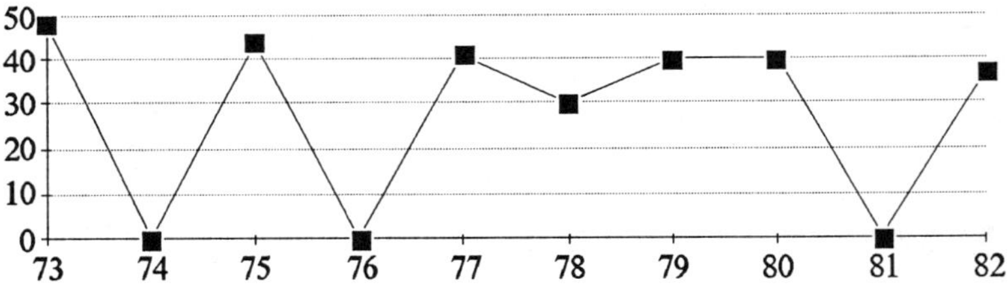

U.S. Percent of the World's S&T Literature:
Percent for Physics, by Year
(No data for empty years)

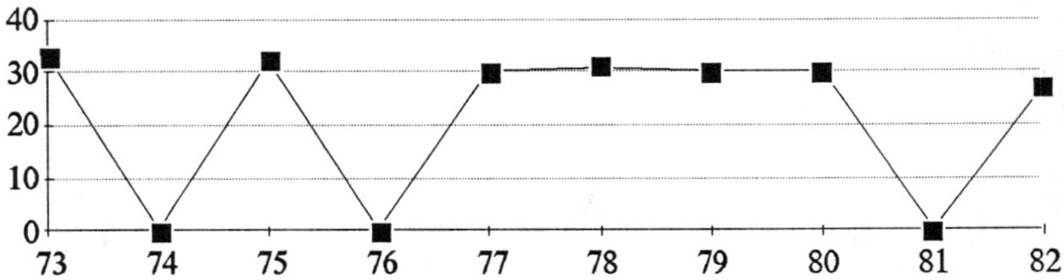

Appendix 3

BIOSIS Records

The following charts show the trends found for languages in the BIOSIS file. The English charts are presented first, followed by other languages in alphabetical order.

Percentages are whole percents: that is, ".65%" indicates sixty-five hundredths of a percent, "1.5%" indicates one and a half percent.

BIOSIS Records for English Documents:
Records by Year

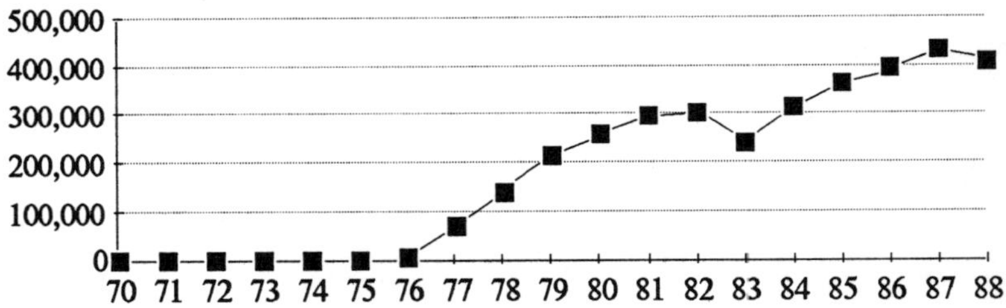

BIOSIS Records for English Documents:
Percent of All Records, by Year

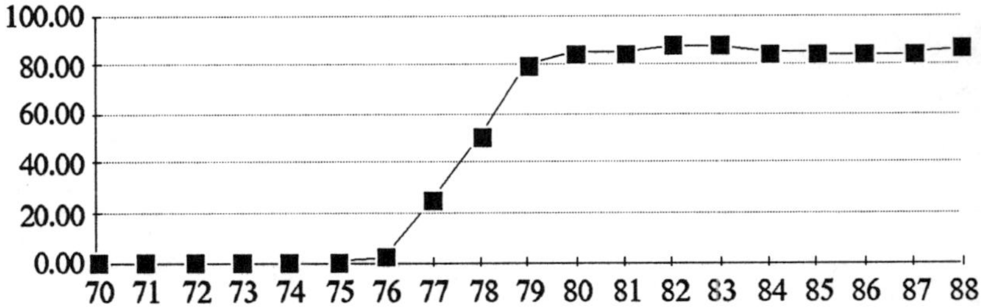

BIOSIS Records for Bulgarian Documents:
Records by Year

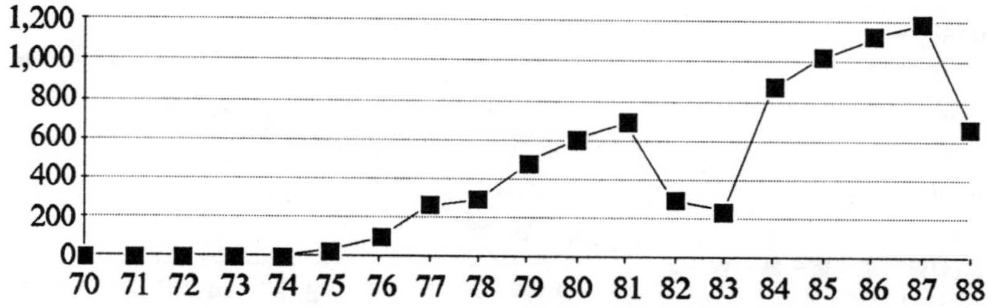

BIOSIS Records for Bulgarian Documents:
Percent of All Records, by Year

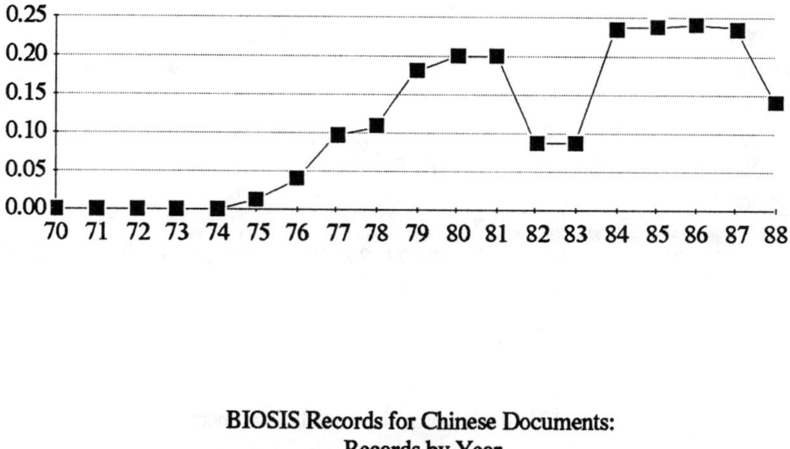

BIOSIS Records for Chinese Documents:
Records by Year

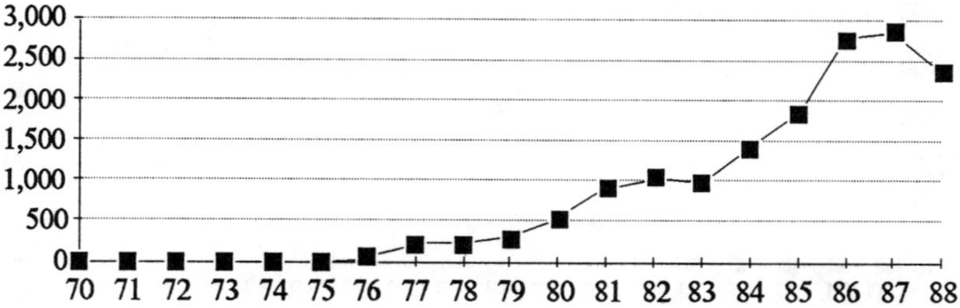

BIOSIS Records for Chinese Documents:
Percent of All Records, by Year

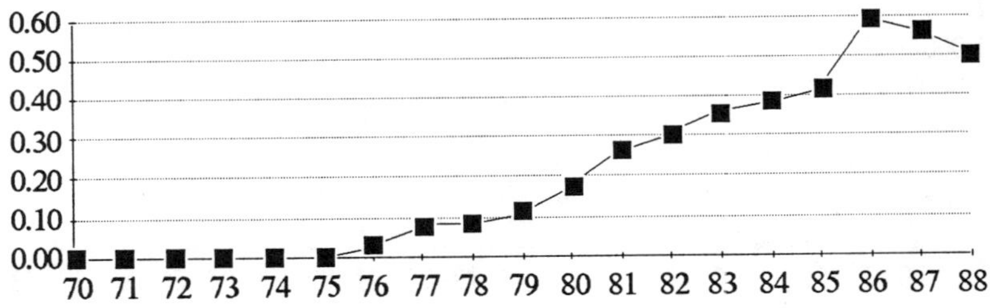

BIOSIS Records for French Documents:
Records by Year

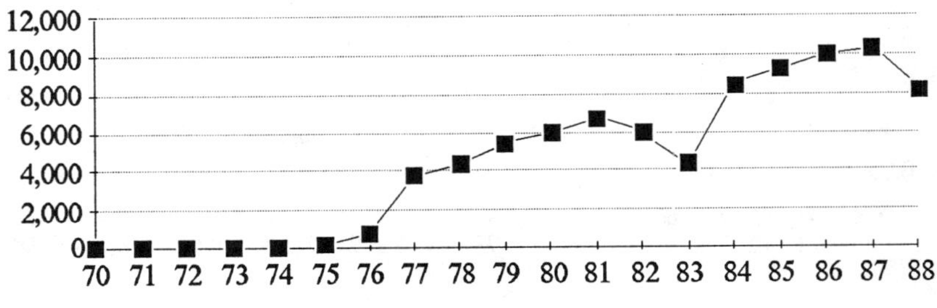

BIOSIS Records for French Documents:
Percent of All Records, by Year

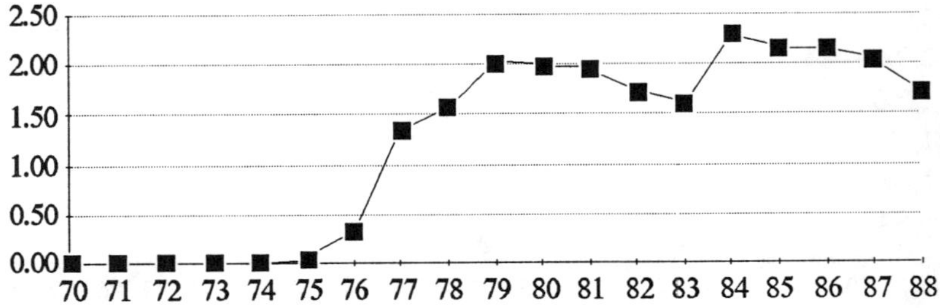

BIOSIS Records for German Documents:
Records by Year

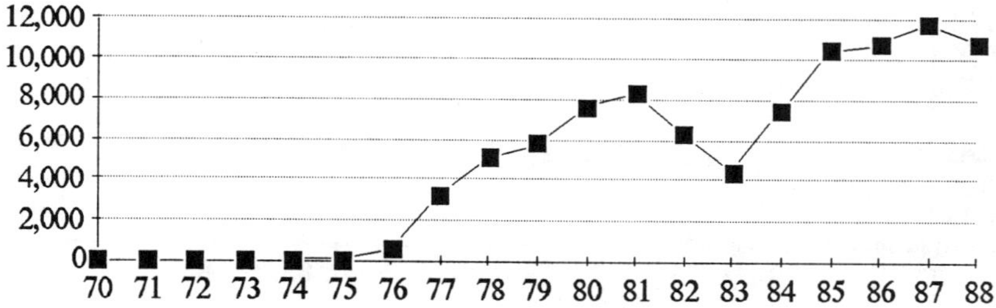

BIOSIS Records for German Documents:
Percent of All Records, by Year

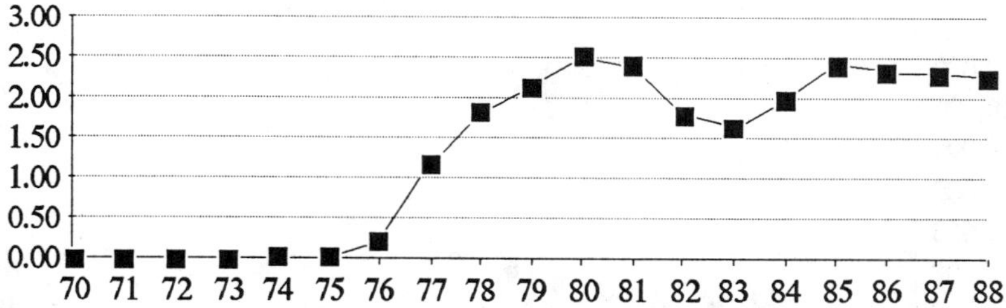

BIOSIS Records for Italian Documents:
Records by Year

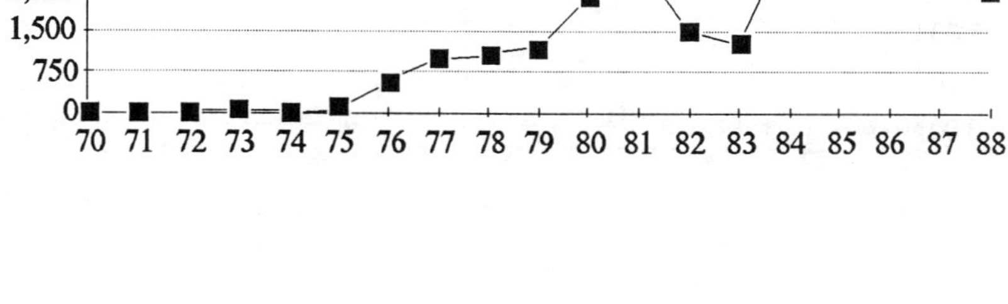

BIOSIS Records for Italian Documents:
Percent of All Records, by Year

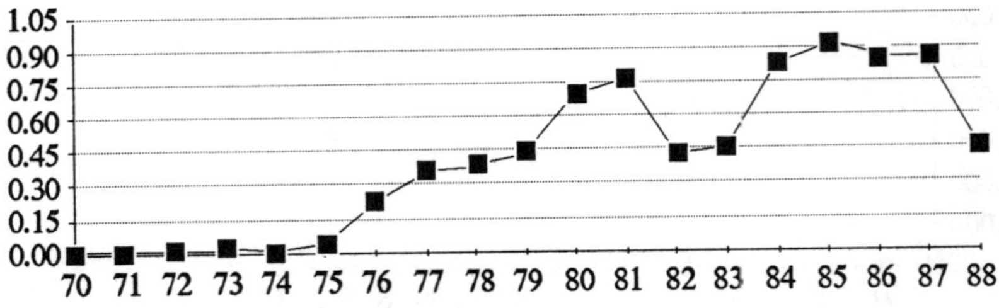

BIOSIS Records for Japanese Documents:
Records by Year

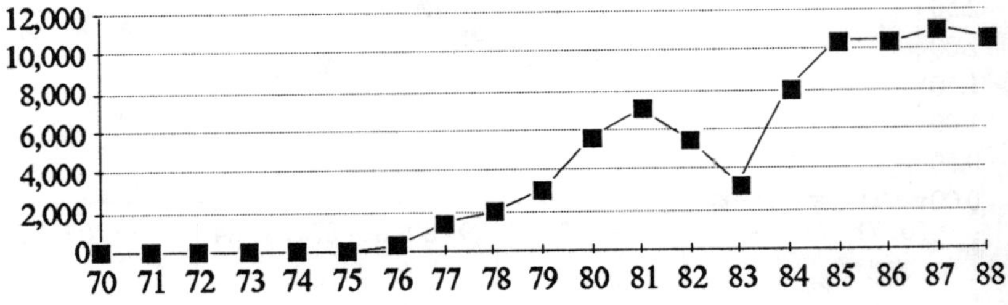

BIOSIS Records for Japanese Documents:
Percent of All Records, by Year

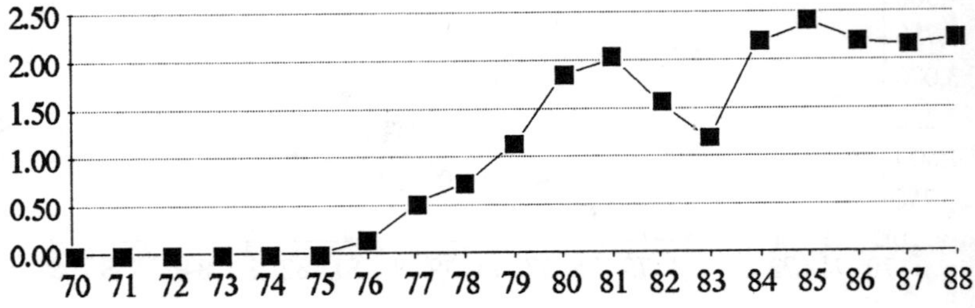

BIOSIS Records for Korean Documents:
Records by Year

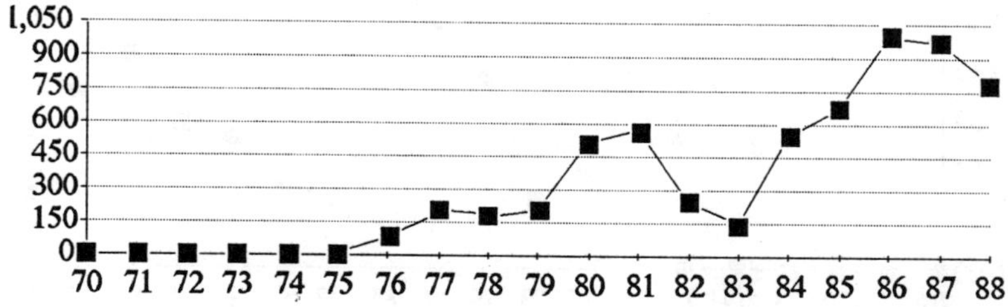

BIOSIS Records for Korean Documents:
Percent of All Records, by Year

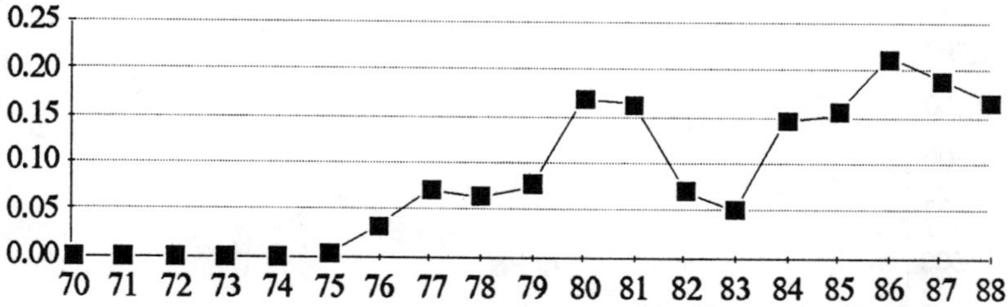

BIOSIS Records for Polish Documents:
Records by Year

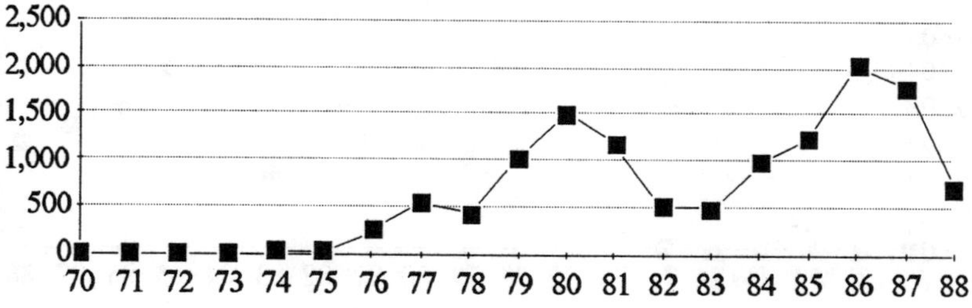

BIOSIS Records for Polish Documents:
Percent of All Records, by Year

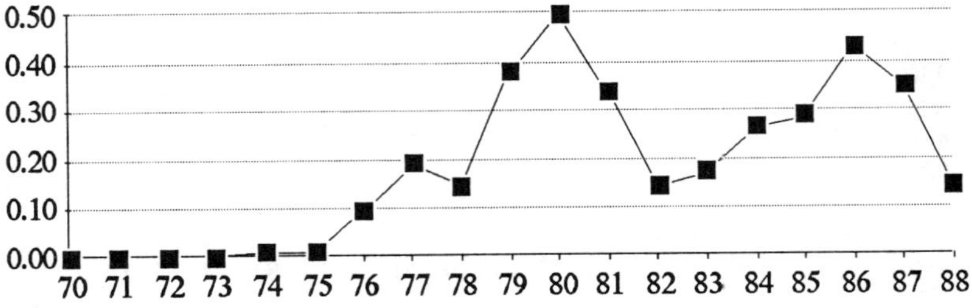

BIOSIS Records for Portuguese Documents:
Records by Year

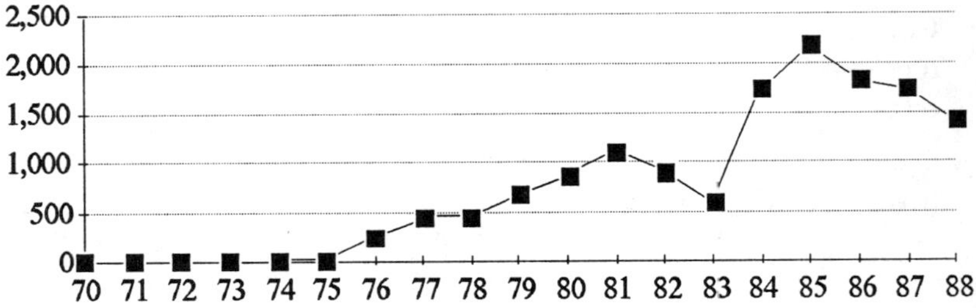

BIOSIS Records for Portuguese Documents:
Percent of All Records, by Year

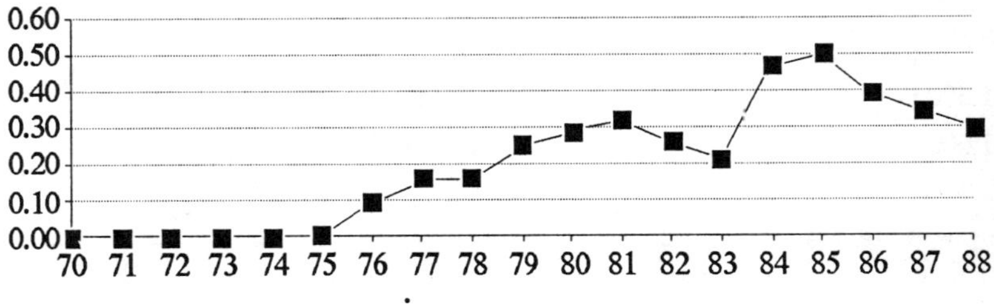

BIOSIS Records for Russian Documents:
Records Per Year

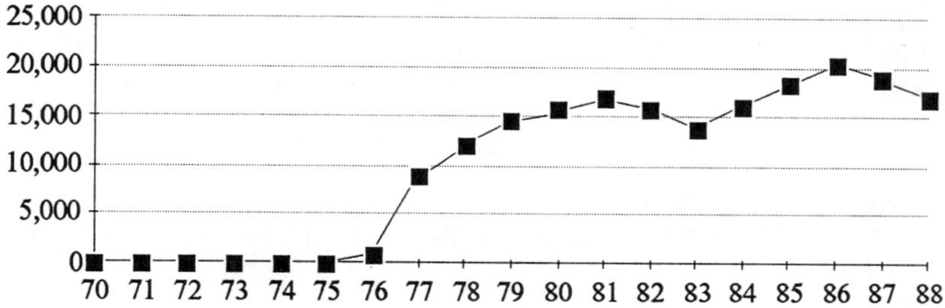

BIOSIS Records for Russian Documents:
Percent of All Records, by Year

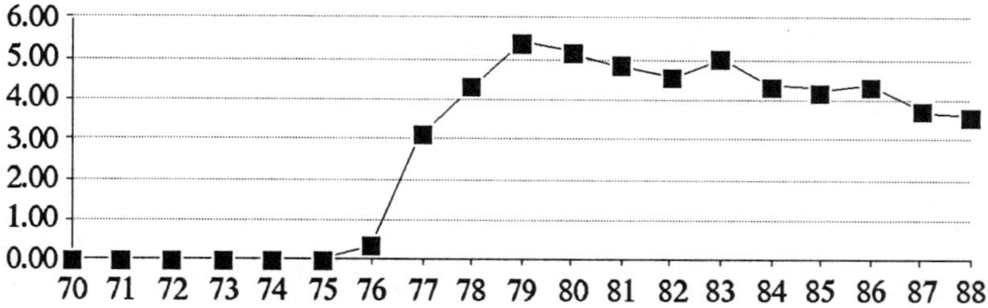

BIOSIS Records for Spanish Documents:
Records by Year

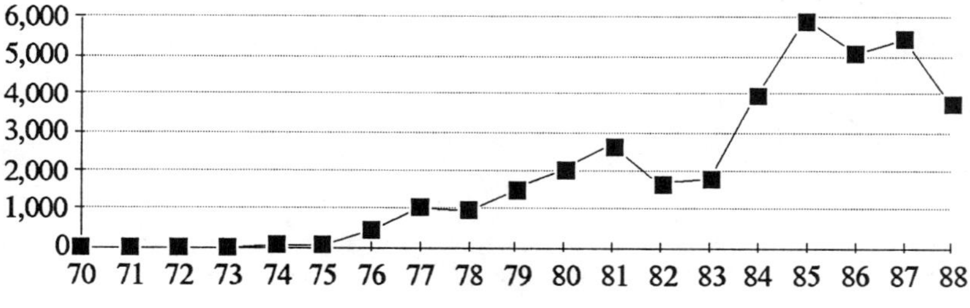

BIOSIS Records for Spanish Documents:
Percent of All Records, by Year

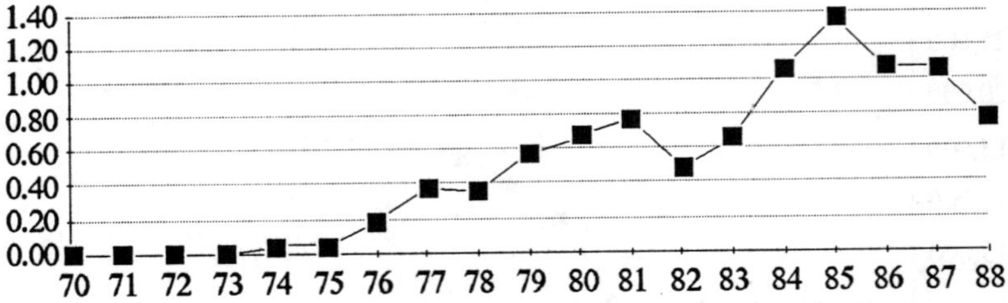

Chemical Abstracts Records

The following charts show the trends found for languages in the Chemical Abstracts file. The English charts are presented first, followed by other languages in alphabetical order.

Percentages are whole percents: that is, ".65%" indicates sixty-five hundredths of a percent, "1.5%" indicates one and a half percent.

Chemical Abstracts Records for English Documents:
Records by Year

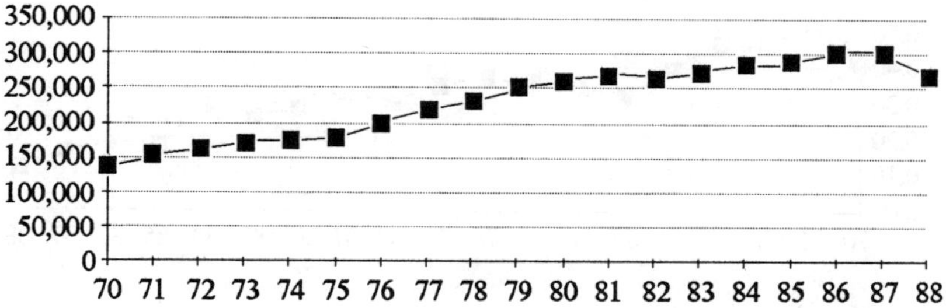

Chemical Abstracts Records for English Documents:
Percent of All Records, by Year

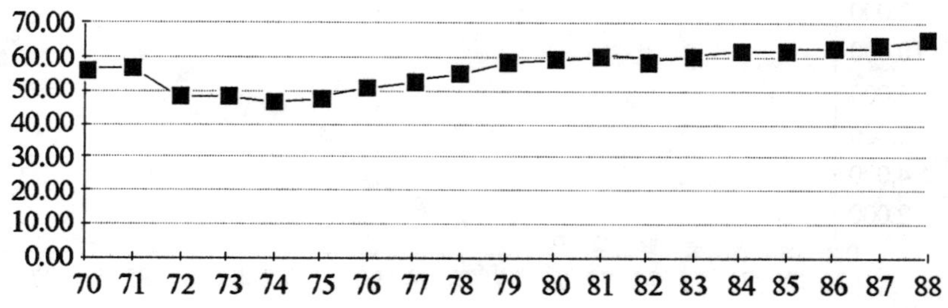

Chemical Abstracts Records for Bulgarian Documents:
Records by Year

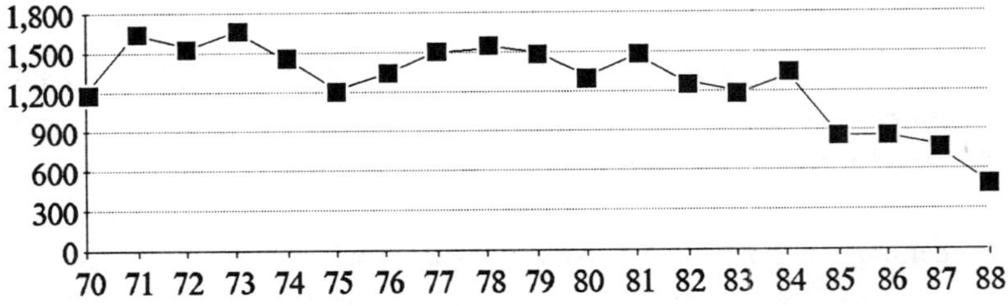

Chemical Abstracts Records for Bulgarian Documents:
Percent of All Records, by Year

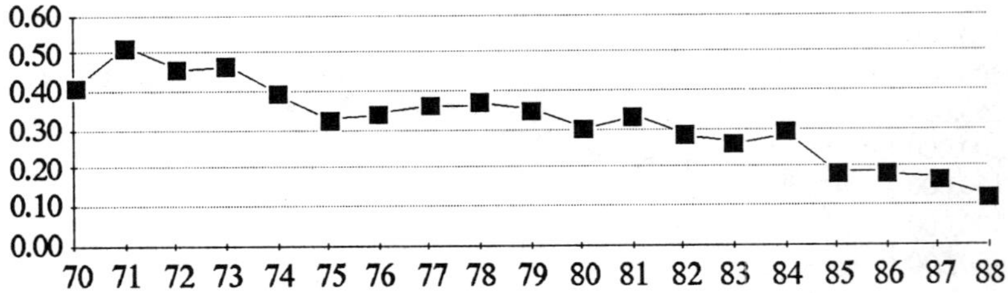

Chemical Abstracts Records for Chinese Documents:
Records by Year

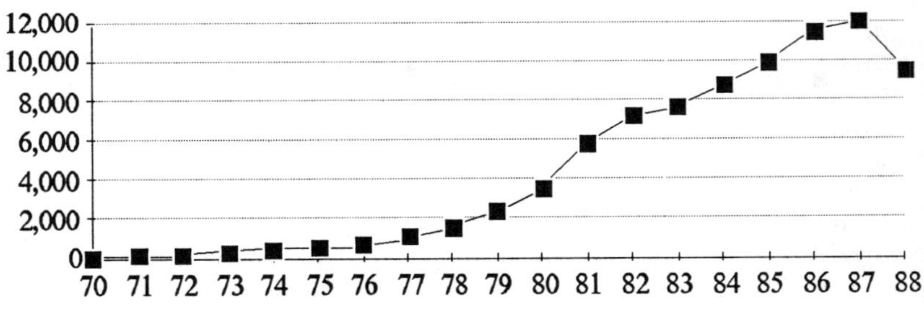

Chemical Abstracts Records for Chinese Documents:
Percent of All Records, by Year

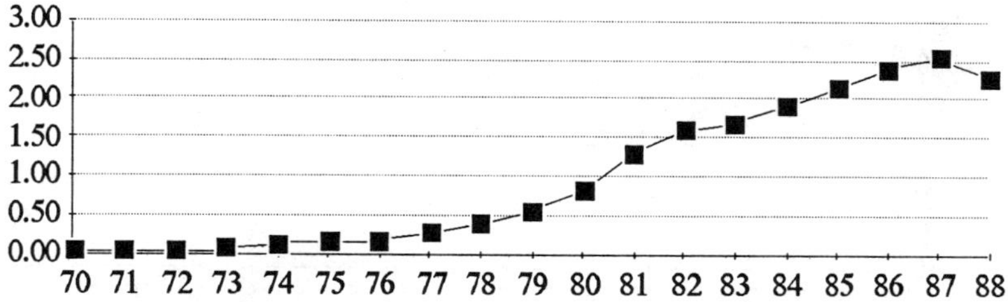

Chemical Abstracts Records for Czech Documents:
Records by Year

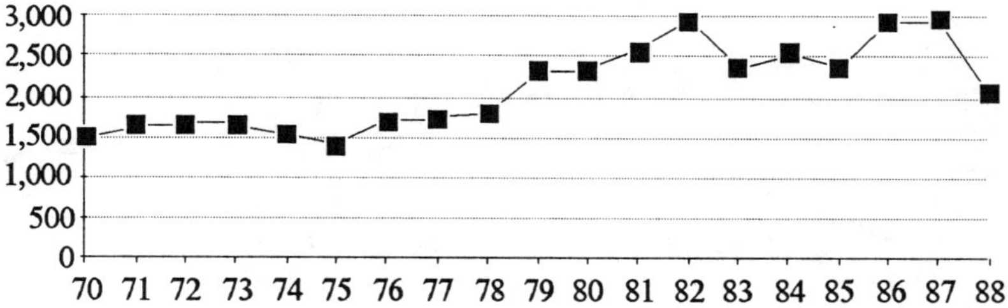

Chemical Abstracts Records for Czech Documents:
Percent of All Records, by Year

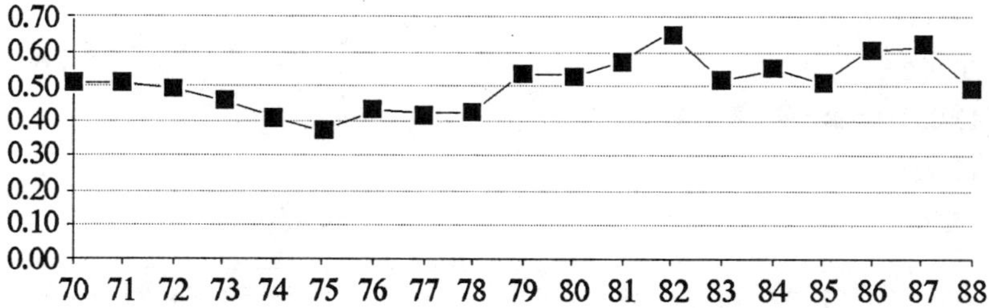

Chemical Abstracts Records for French Documents:
Records by Year

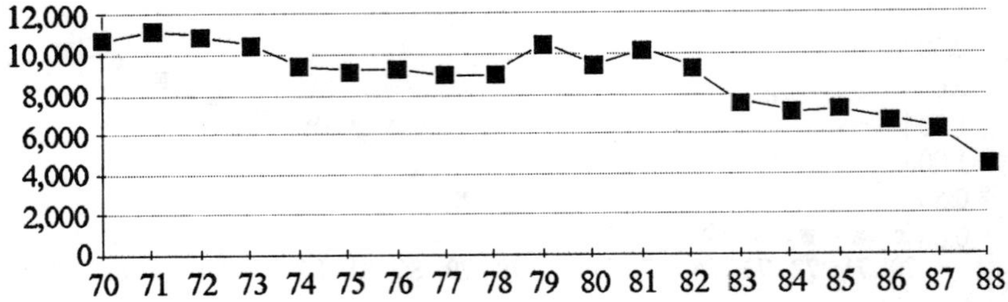

Chemical Abstracts Records for French Documents:
Percent of All Records, by Year

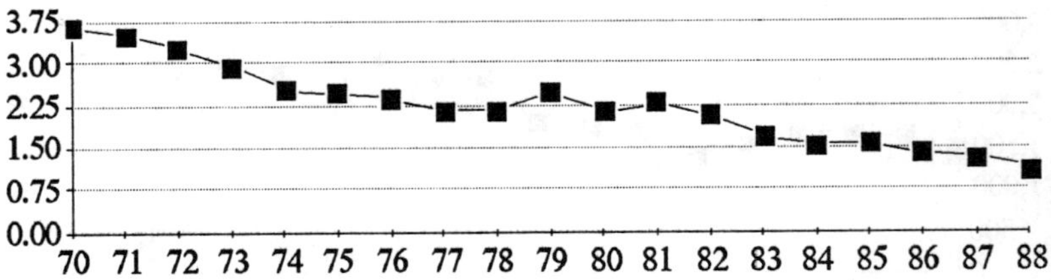

Chemical Abstracts Records for German Documents:
Records by Year

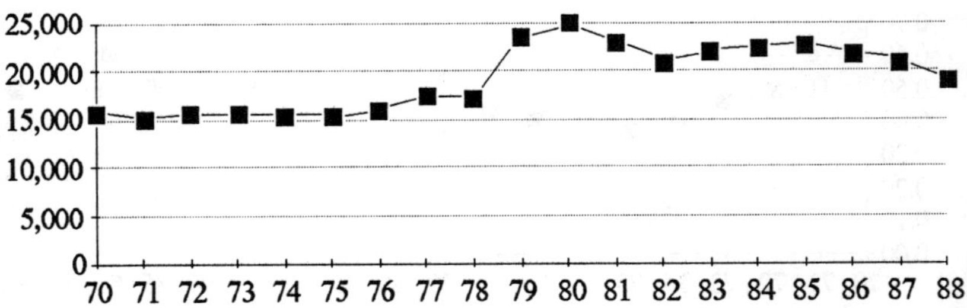

Chemical Abstracts Records for German Documents:
Percent of All Records, by Year

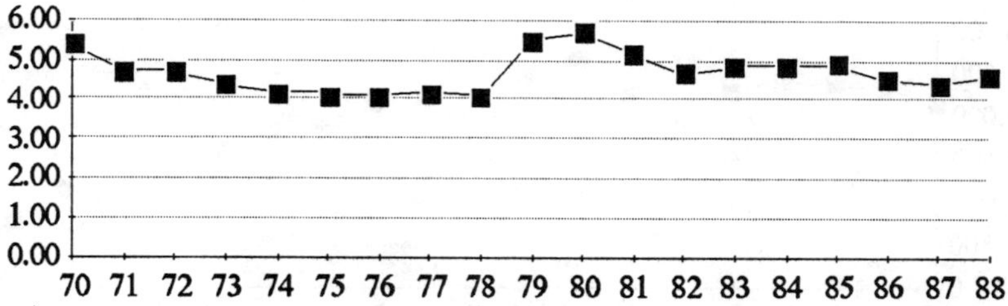

Chemical Abstracts Records for Hungarian Documents:
Records by Year

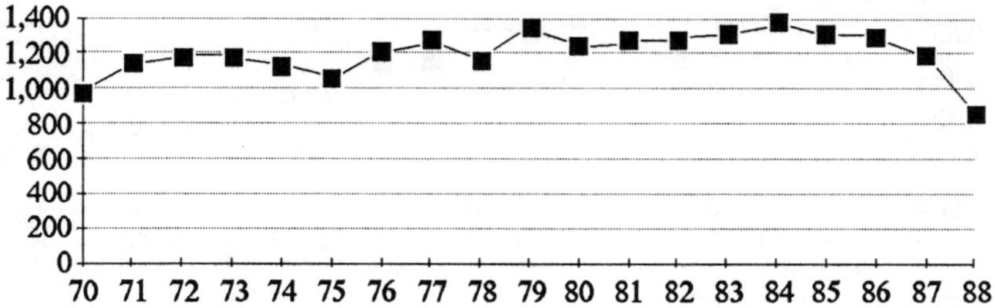

Chemical Abstracts Records for Hungarian Documents:
Percent of All Records, by Year

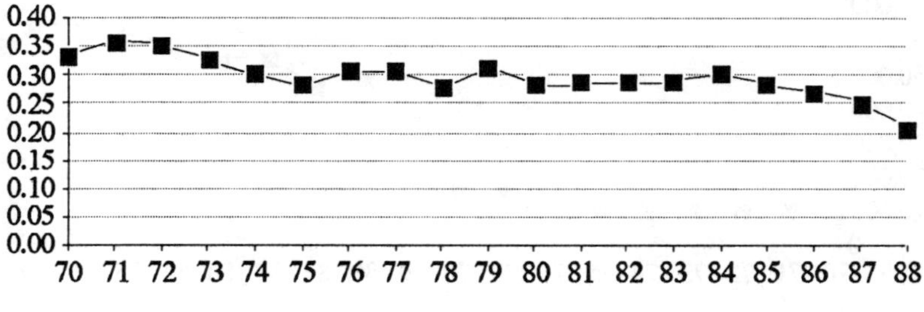

Chemical Abstracts Records for Italian Documents:
Records by Year

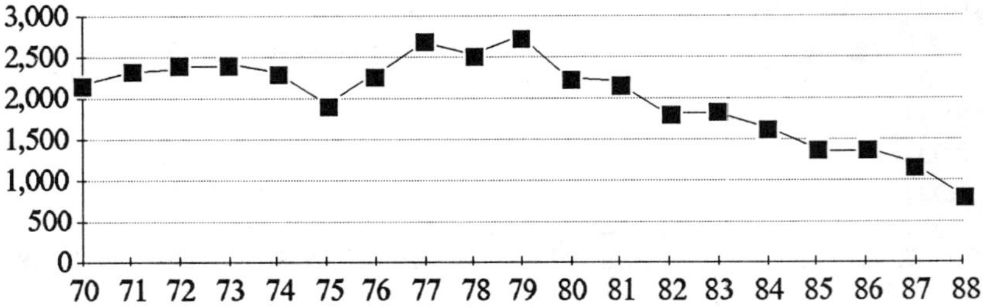

Chemical Abstracts Records for Italian Documents:
Percent of All Records, by Year

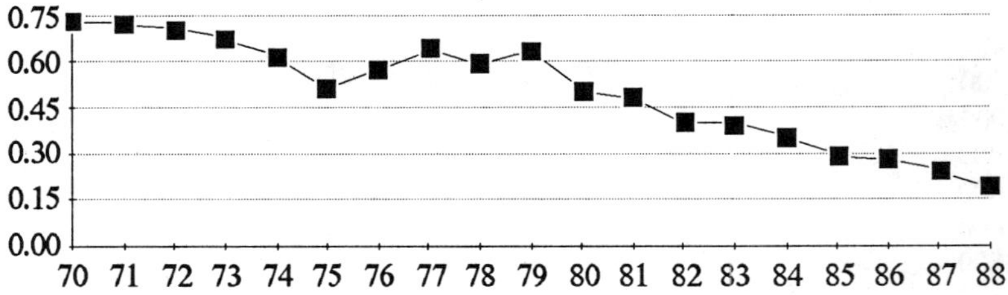

Chemical Abstracts Records for Japanese Documents:
Records by Year

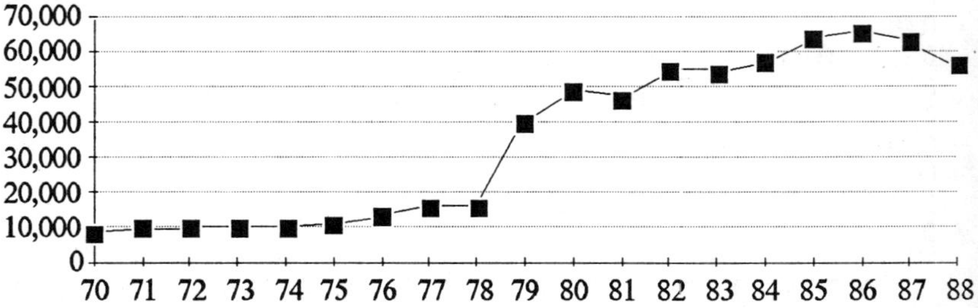

Chemical Abstracts Records for Japanese Documents:
Percent of All Records, by Year

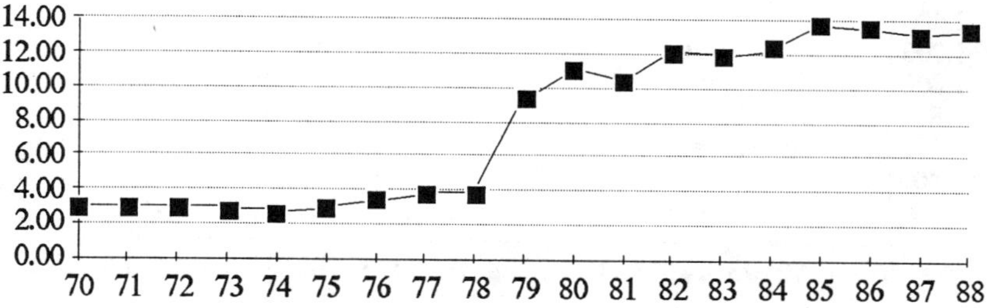

Chemical Abstracts Records for Polish Documents:
Records by Year

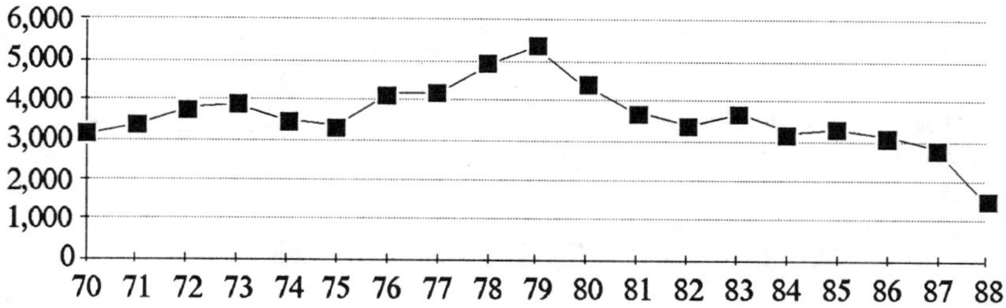

Chemical Abstracts Records for Polish Documents:
Percent of All Records, by Year

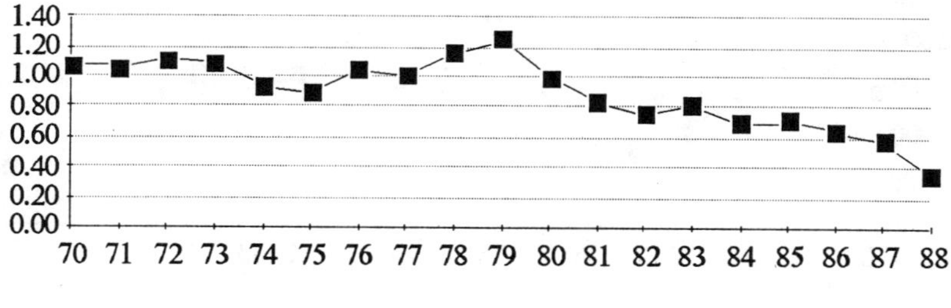

Chemical Abstracts Records for Portuguese Documents:
Records by Year

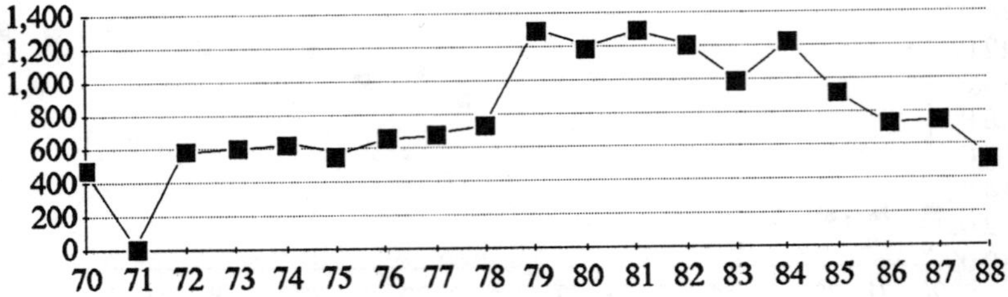

Chemical Abstracts Records for Portuguese Documents:
Percent of All Records, by Year

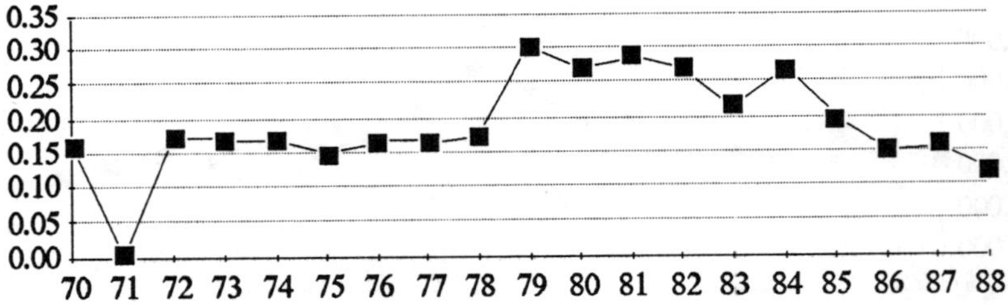

Chemical Abstracts Records for Romanian Documents:
Records by Year

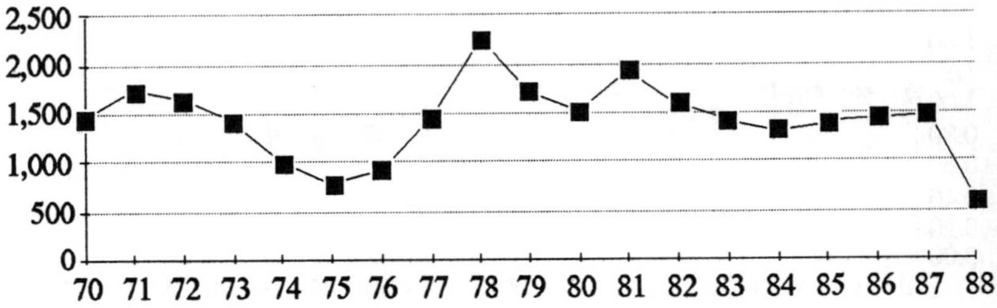

Chemical Abstracts Records for Romanian Documents:
Percent of All Records, by Year

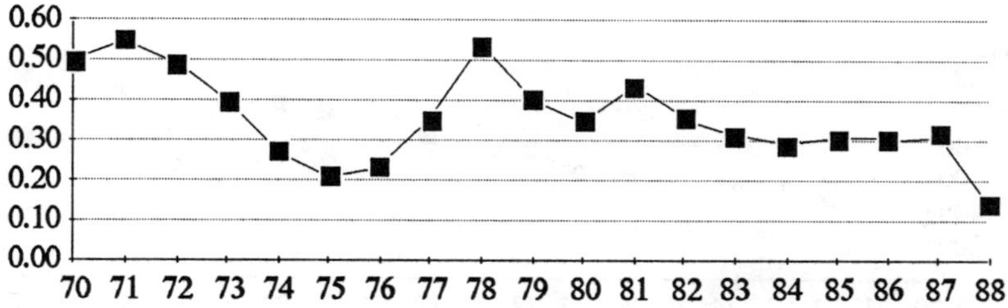

Chemical Abstracts Records for Russian Documents:
Records by Year

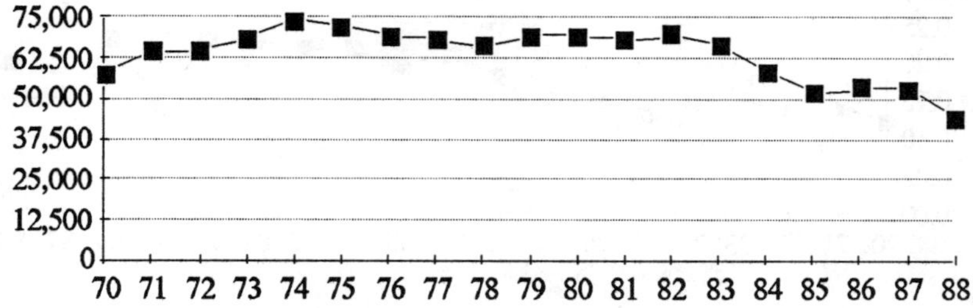

Chemical Abstracts Records for Russian Documents:
Percent of All Records, by Year

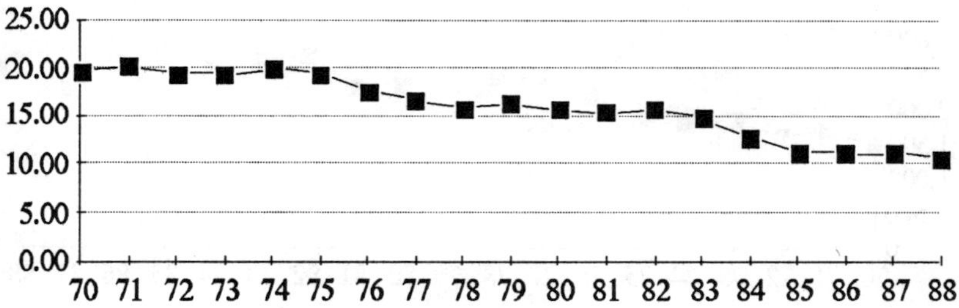

Chemical Abstracts Records for Slovak Documents:
Records by Year

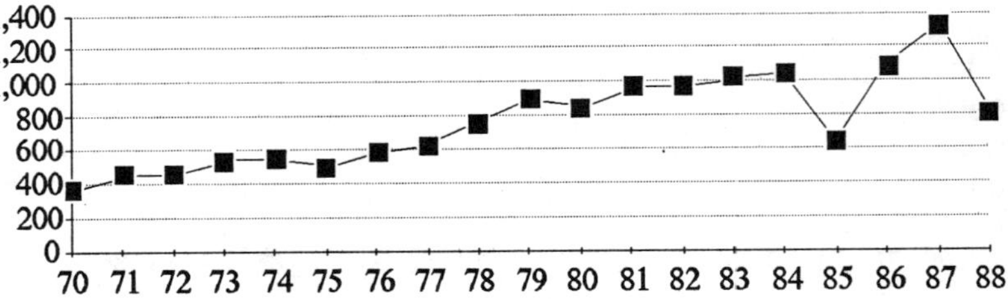

Chemical Abstracts Records for Slovak Documents:
Percent of All Records, by Year

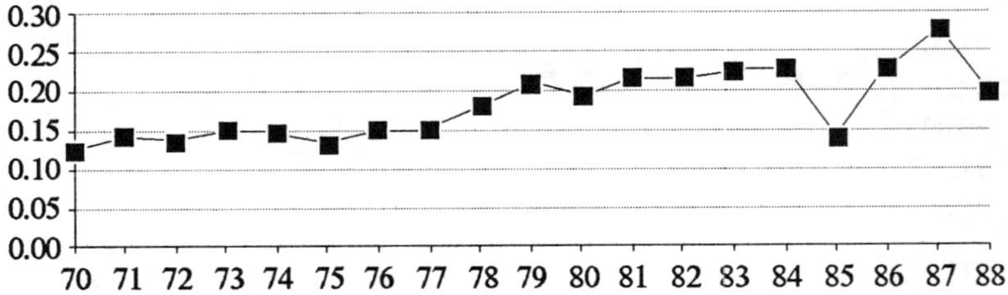

Chemical Abstracts Records for Spanish Documents:
Records by Year

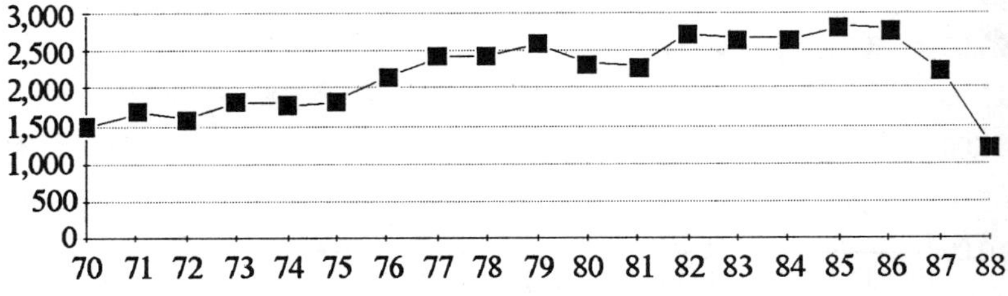

Chemical Abstracts Spanish Records:
Percent of All Records, by Year

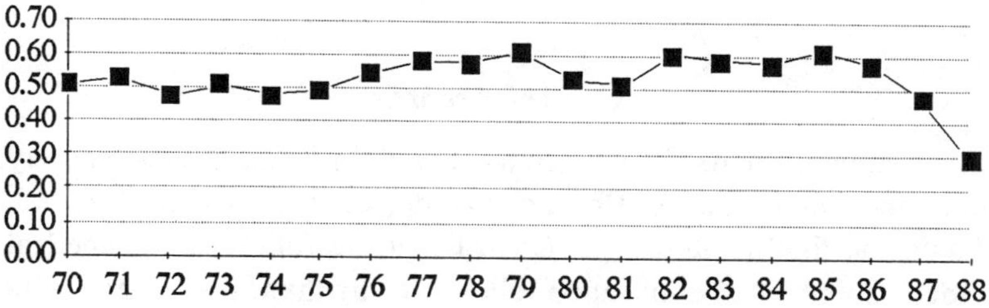

Chemical Abstracts Records for Ukrainian Documents:
Records by Year

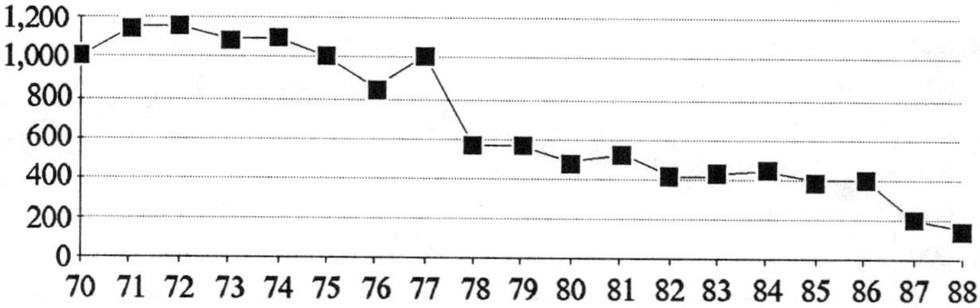

Chemical Abstracts Records for Ukrainian Documents:
Percent of Total Records, by Year

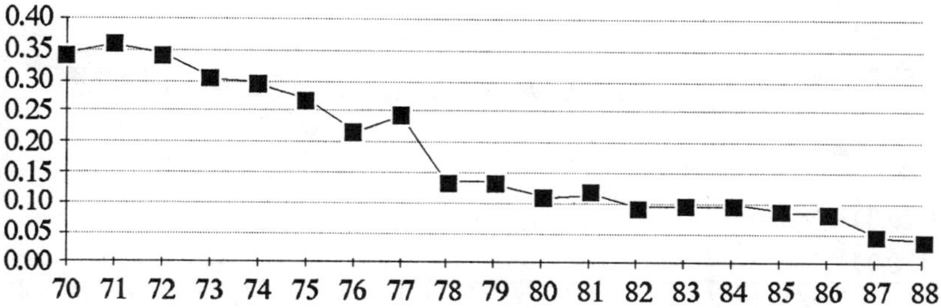

GeoRef Records

The following charts show the trends found for languages and country of publication for the GeoRef file. Languages are presented first, with English as the first language. Other languages follow in alphabetical order. Countries of publication follow, led by the United States, with other countries following in alphabetical order. Percentages are whole percents: That is, ".65%" indicates sixty-five hundredths of a percent, "1.5%" indicates one and a half percent.

GeoRef Records for English Documents:
Records by Year

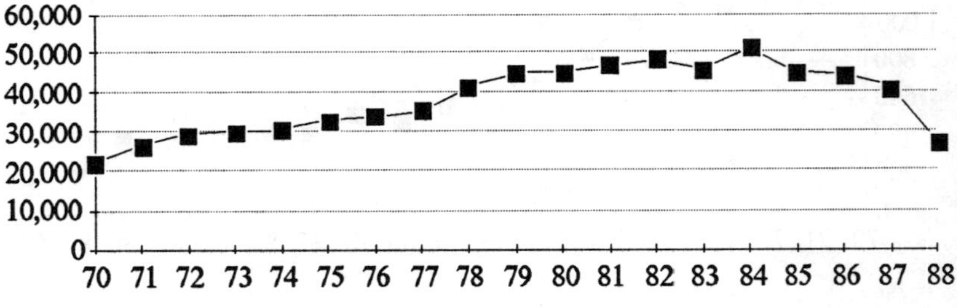

GeoRef Records for English Documents:
Percent of All Records, by Year

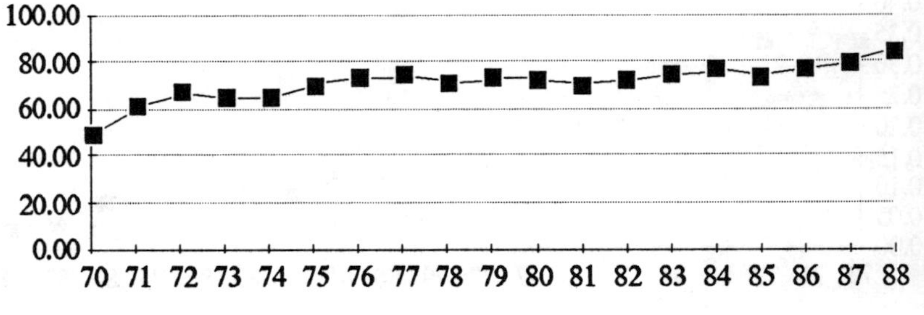

GeoRef Records for Chinese Documents:
Records by Year

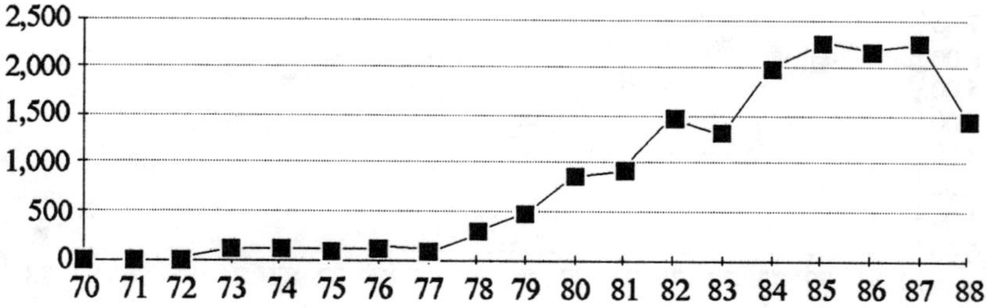

GeoRef Records for Chinese Documents:
Percent of All Records, by Year

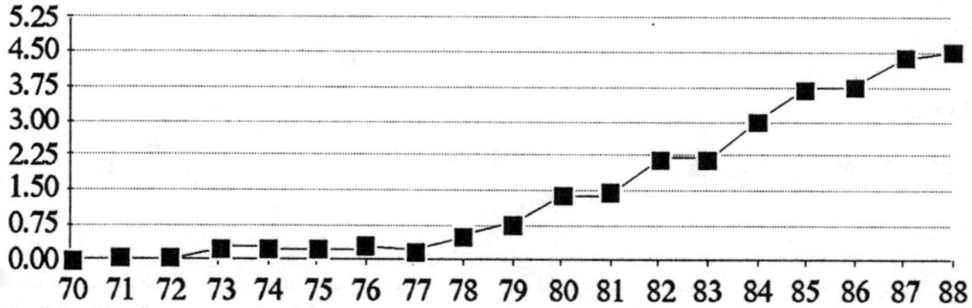

GeoRef Records for French Documents:
Records by Year

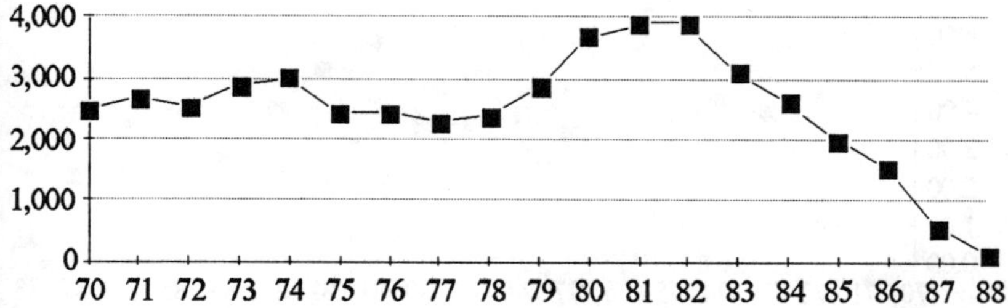

GeoRef Records for French Documents:
Percent of All Records, by Year

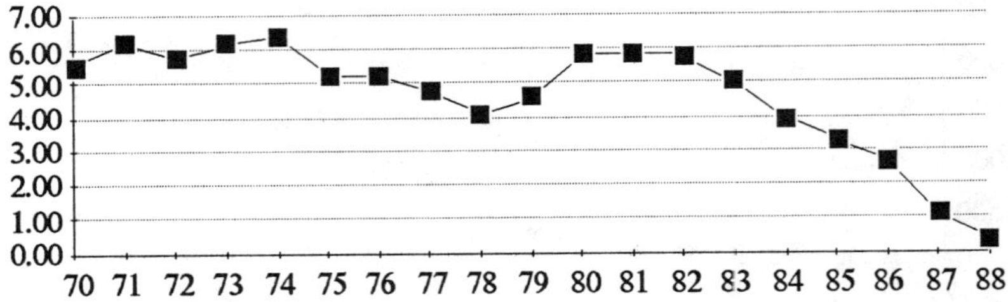

GeoRef Records for German Documents:
Records by Year

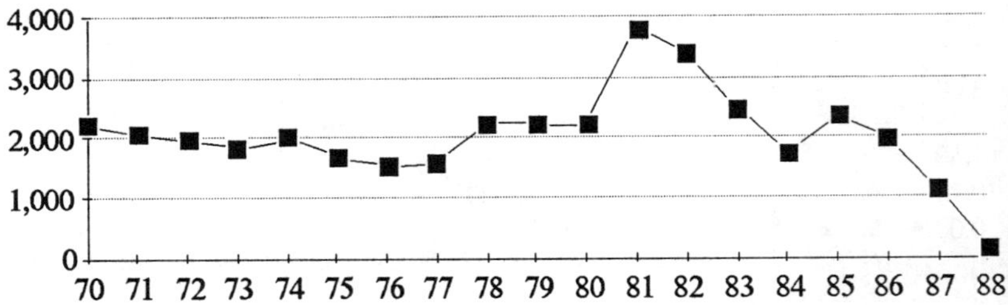

GeoRef Records for German Documents:
Percent of All Records, by Year

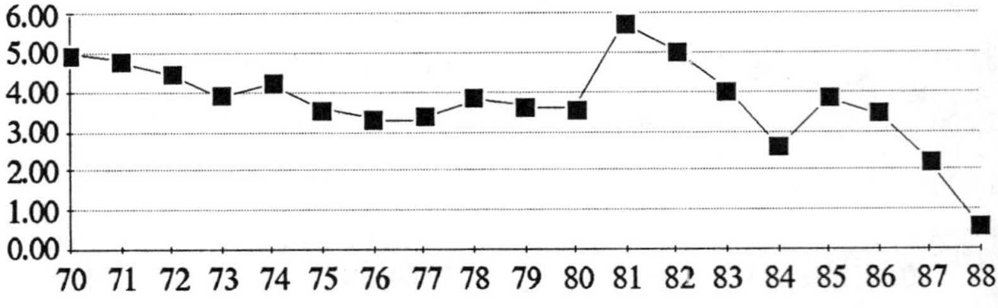

GeoRef Records for Japanese Documents:
Records by Year

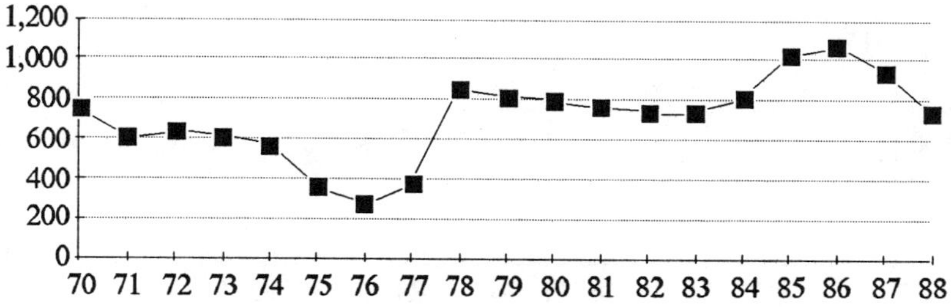

GeoRef Records for Japanese Documents:
Percent of All Records, by Year

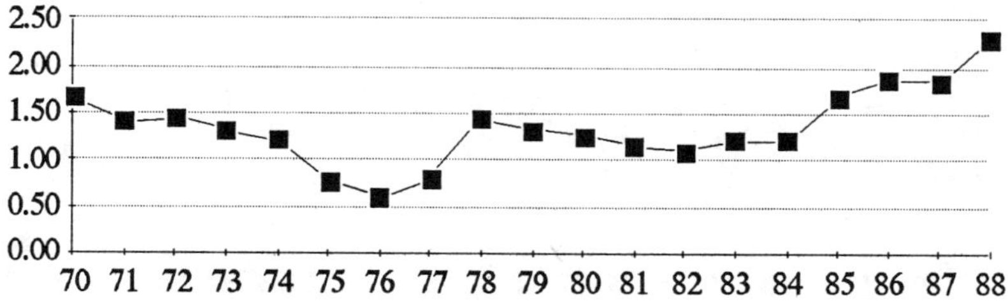

GeoRef Records for Russian Documents:
Records by Year

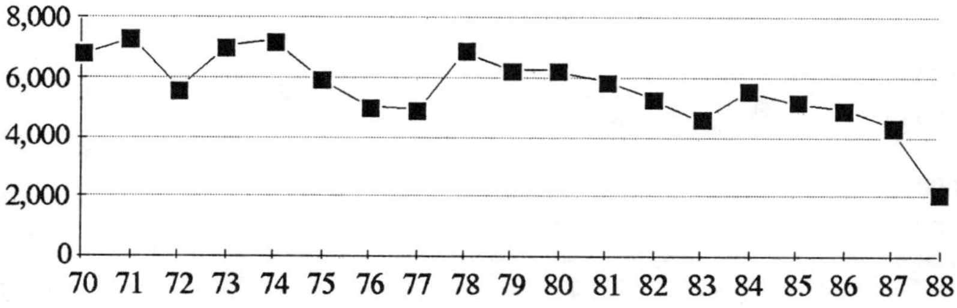

GeoRef Records for Russian Documents:
Percent of All Records, by Year

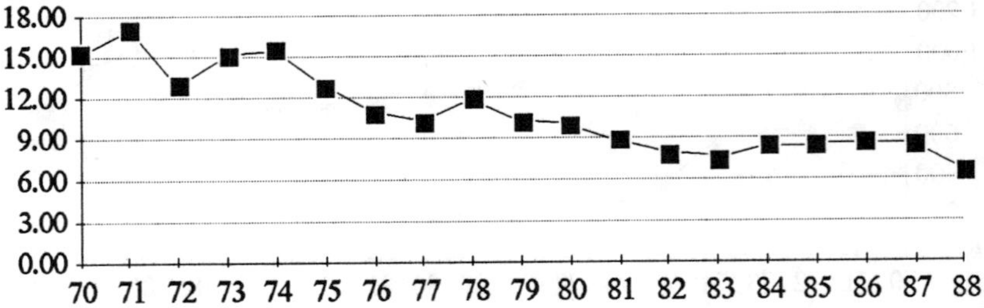

GeoRef Records for Spanish Documents:
Records by Year

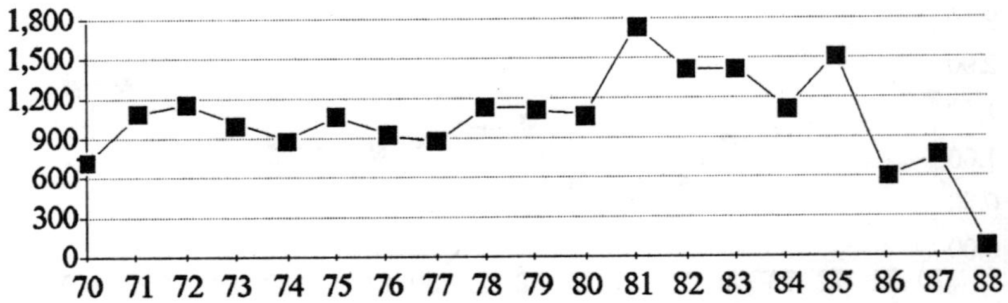

GeoRef Records for Spanish Documents:
Percent of All Records, by Year

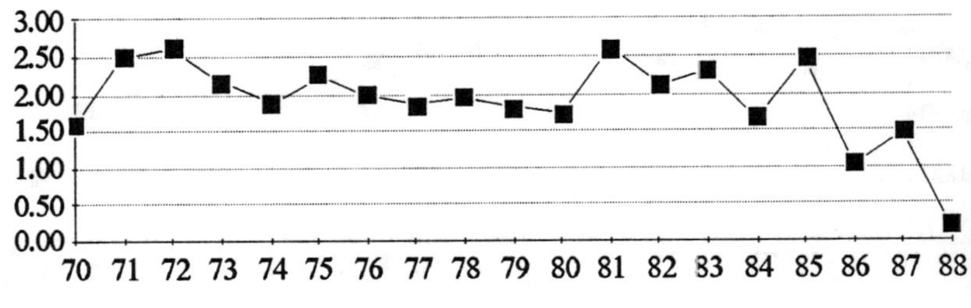

GeoRef Countries of Publication

GeoRef Records for Documents from the United States:
Records by Year

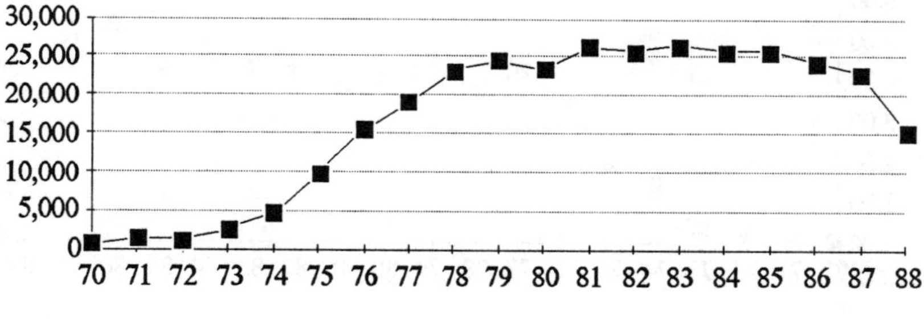

GeoRef Records for Documents from the United States:
Percent of All Records, by Year

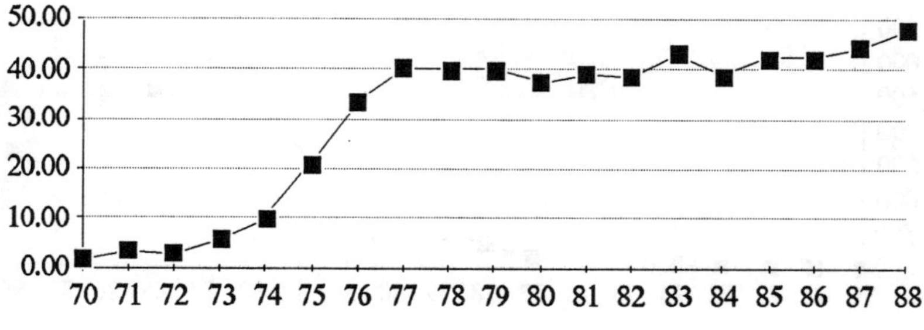

GeoRef Records for Documents from Canada::
Records by Year

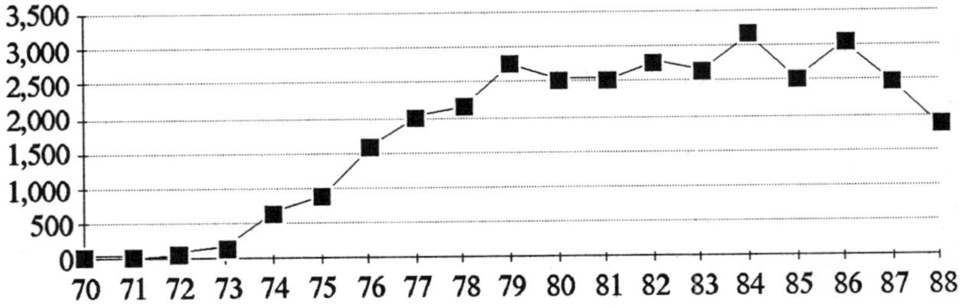

GeoRef Records for Documents from Canada:
Percent of All Records, by Year

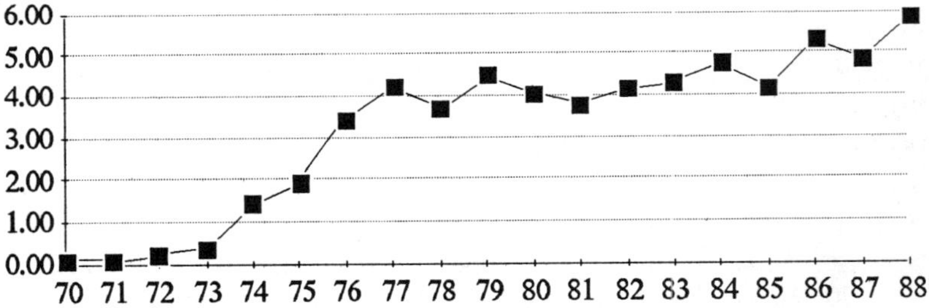

GeoRef Records for Documents from China:
Records by Year

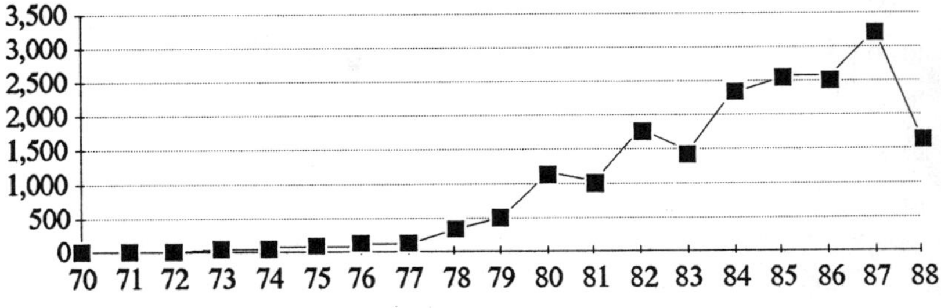

GeoRef Records for Documents from China:
Percent of All Records, by Year

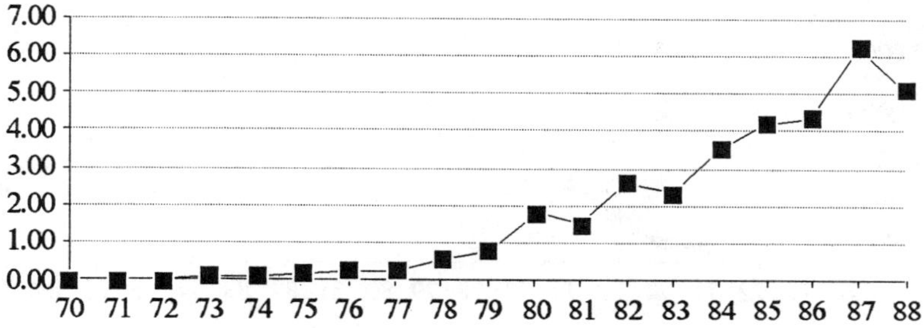

GeoRef Records for Documents from the F. R. Germany: :
Records by Year

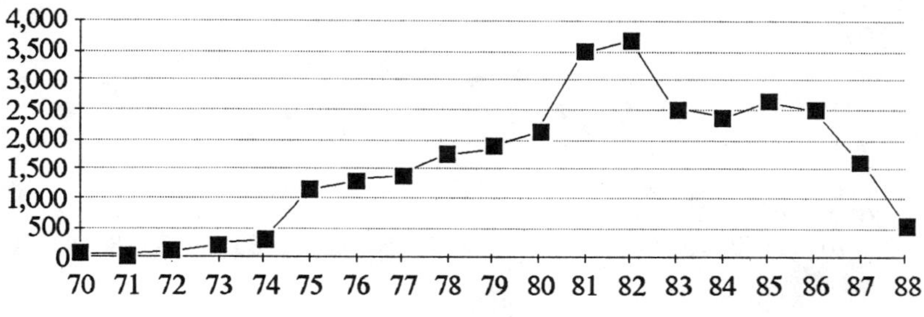

GeoRef Records for Documents from the F. R. Germany:
Percent of All Records, by Year

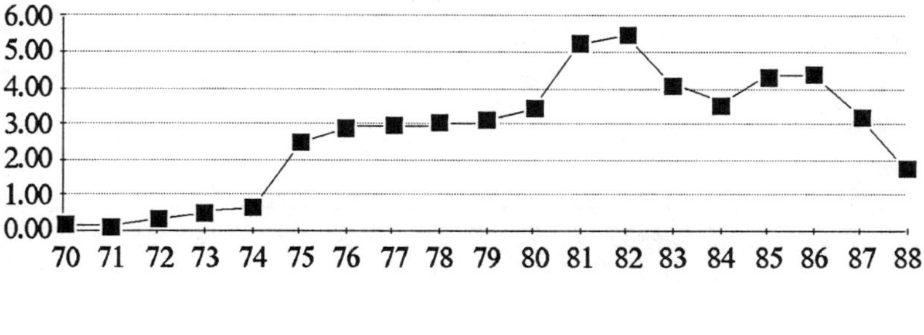

GeoRef Records Coded "International":
Records by Year

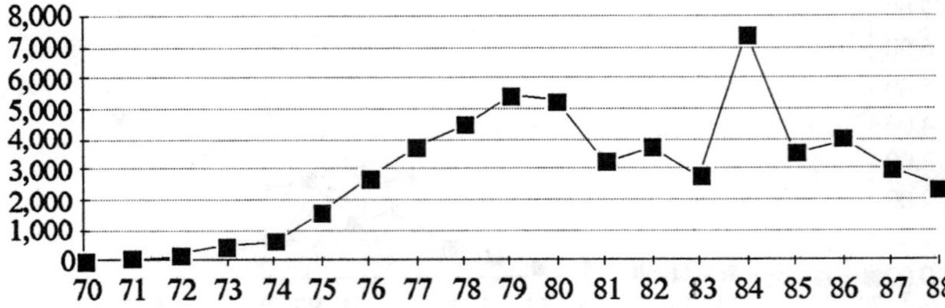

GeoRef Records Coded "International":
Percent of All Records, by Year

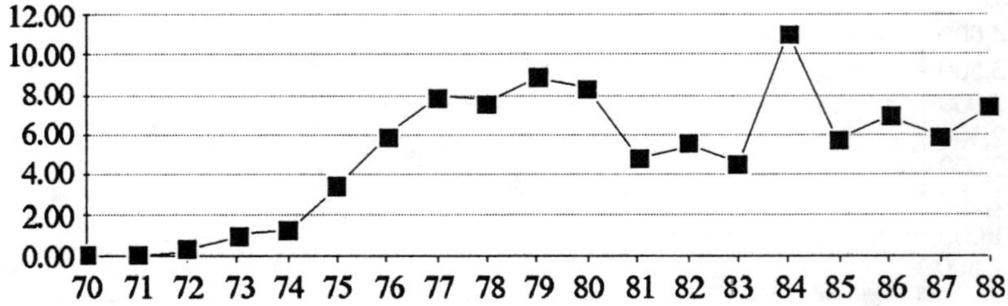

GeoRef Records from Japan:
Records by Year

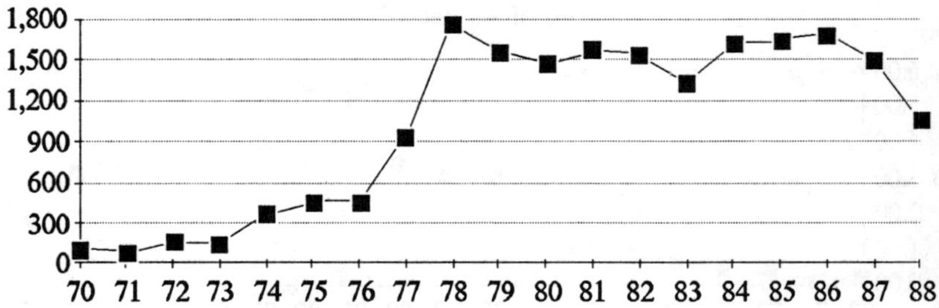

GeoRef Records for Documents from Japan:
Percent of All Records, by Year

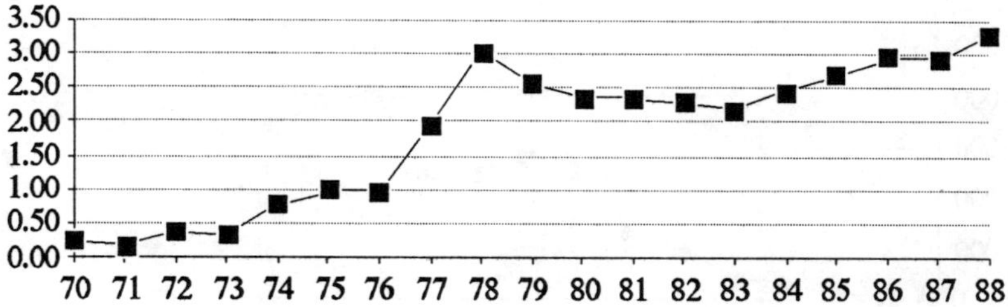

GeoRef Records for Documents from the Netherlands:
Records by Year

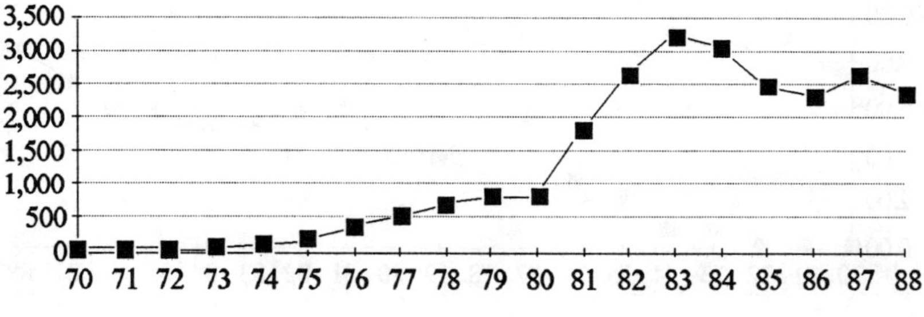

GeoRef Records for Documents from the Netherlands:
Percent of All Records, by Year

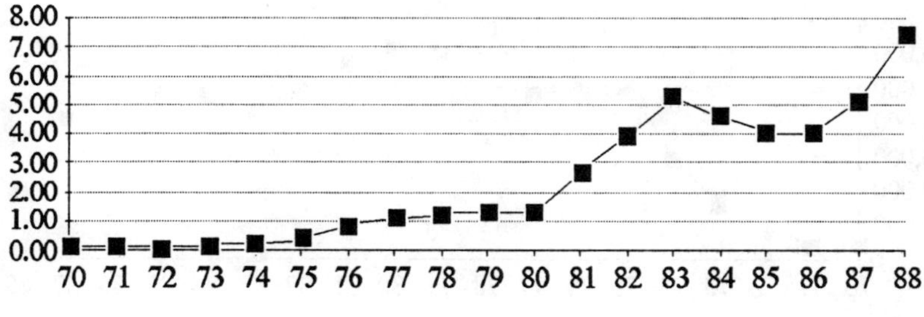

GeoRef Records for Documents from the U. K.:
Records by Year

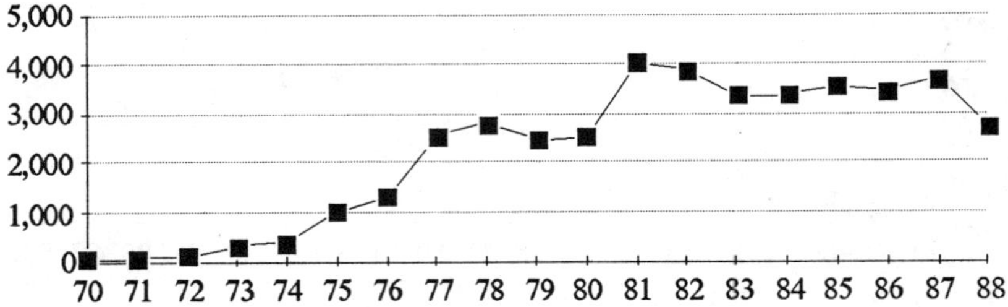

GeoRef Records for Documents from the U. K.:
Percent of All Records, by Year

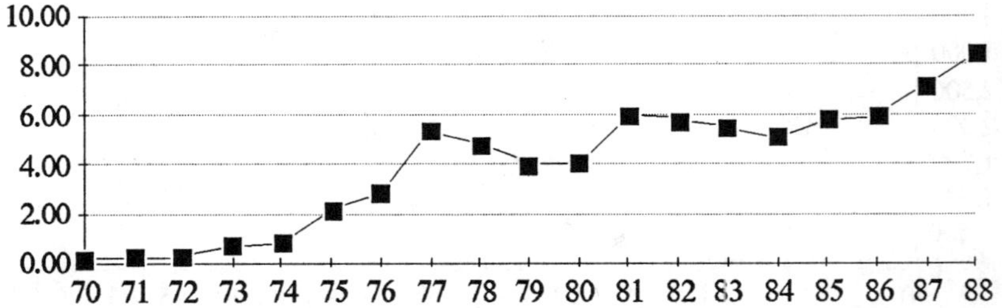

GeoRef Records for Documents from the USSR:
Records by Year

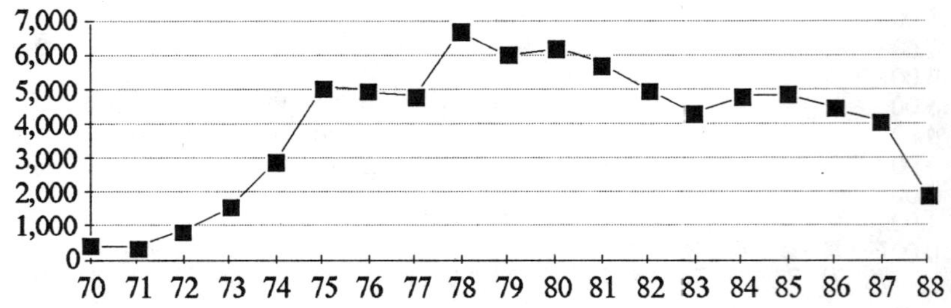

GeoRef Records for Documents from the USSR:
Percent of All Records, by Year

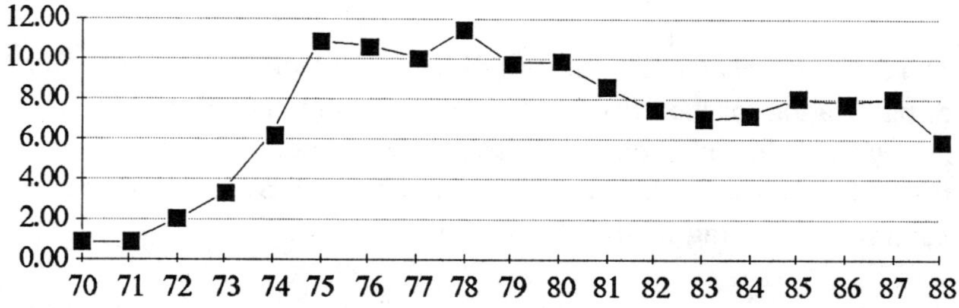

MEDLINE Records

The following charts show the trends found for languages and country of publication for the MEDLINE file. Languages are presented first, led by English. Other languages follow in alphabetical order. Countries of publication follow, led by the United States, with other countries following in alphabetical order.

Percentages are whole percents: that is, ".65%" indicates sixty-five hundredths of a percent, "1.5%" indicates one and a half percent.

MEDLINE Records for English Documents:
Records by Year

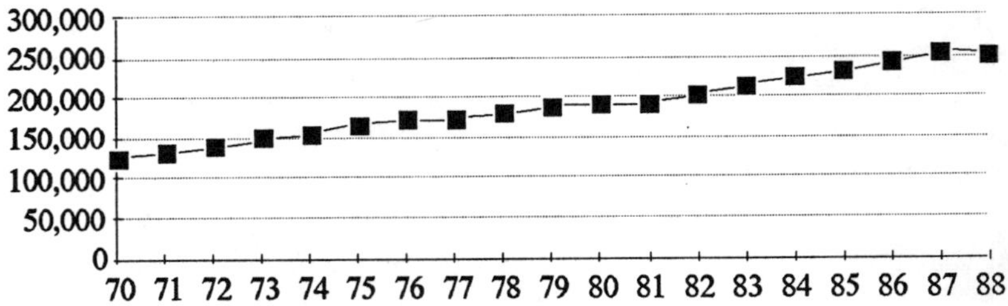

MEDLINE Records for English Documents:
Percent of All Records, by Year

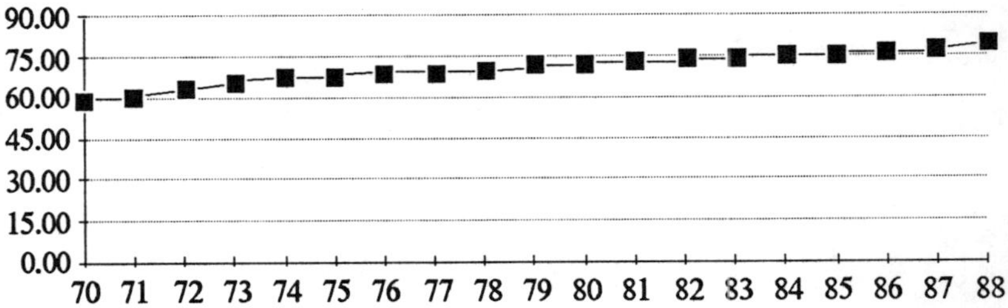

MEDLINE Records for Chinese Documents:
Records by Year

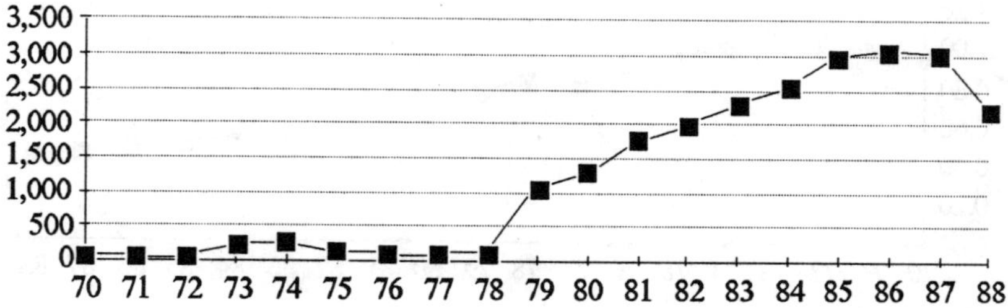

MEDLINE Records for Chinese Documents:
Percent of All Records, by Year

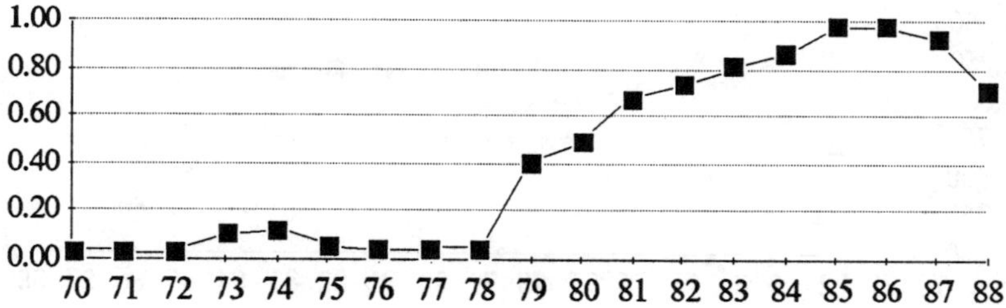

MEDLINE Records for Czech Documents:
Records by Year

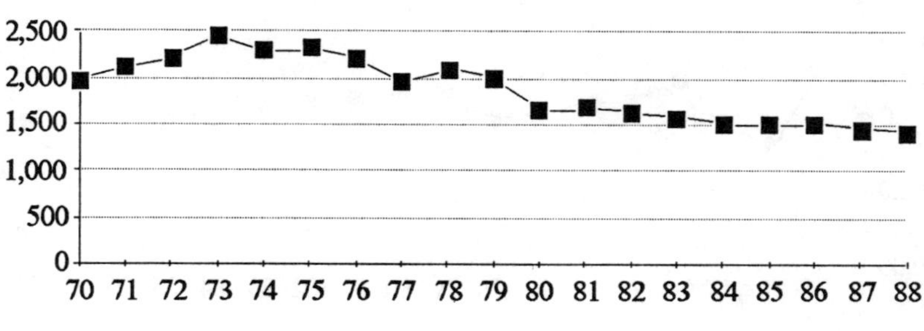

MEDLINE Records for Czech Documents:
Percent of All Records, by Year

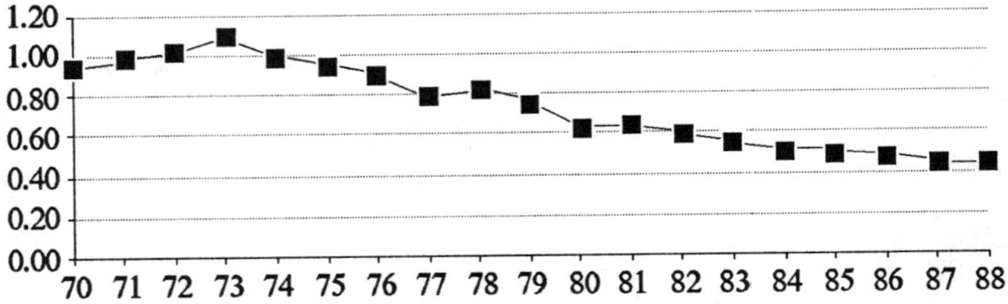

MEDLINE Records for Danish Documents:
Records by Year

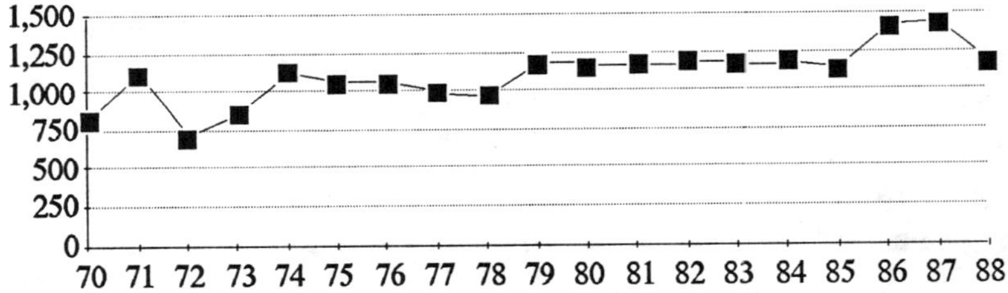

MEDLINE Records for Danish Documents:
Percent of All Records, by Year

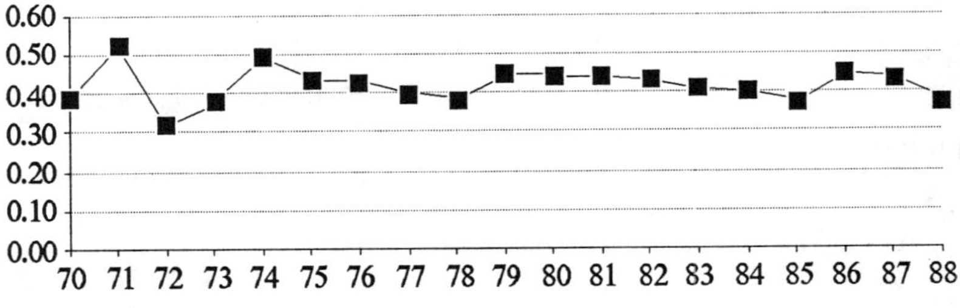

MEDLINE Records for Dutch Documents:
Records by Year

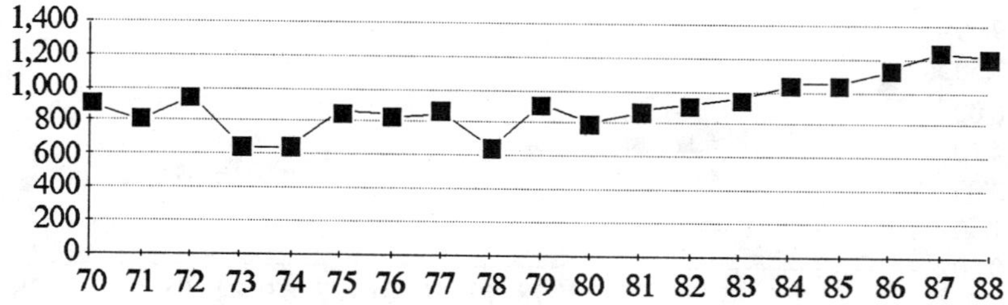

MEDLINE Records for Dutch Documents:
Percent of All Records, by Year

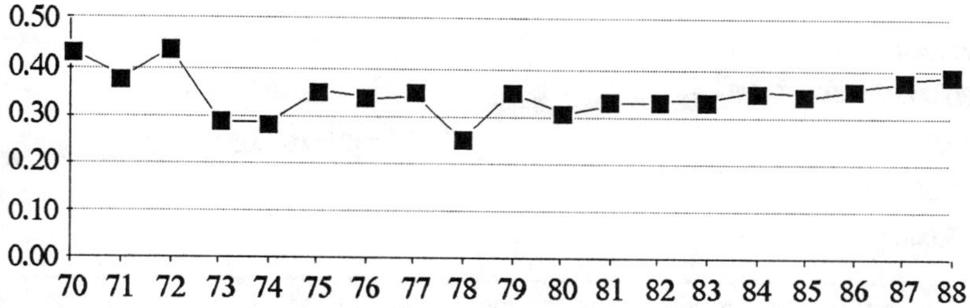

MEDLINE Records for French Documents:
Records by Year

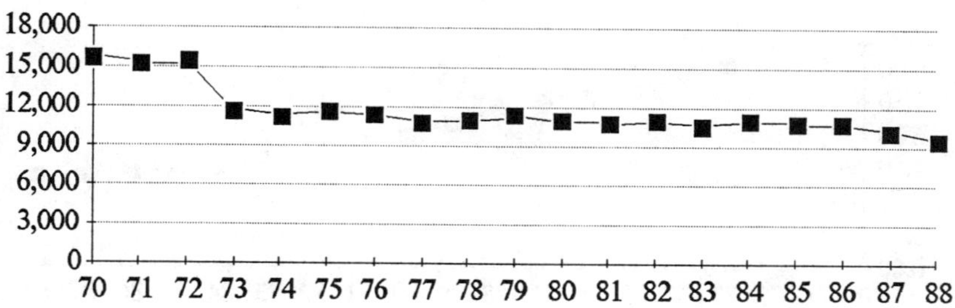

MEDLINE Records for French Documents:
Percent of All Records, by Year

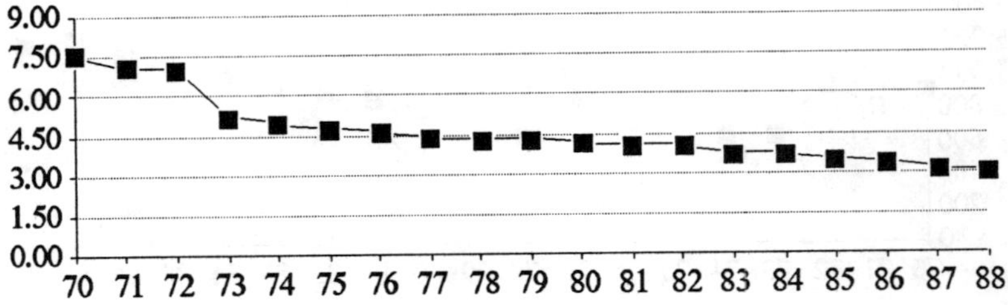

MEDLINE Records for German Documents:
Records by Year

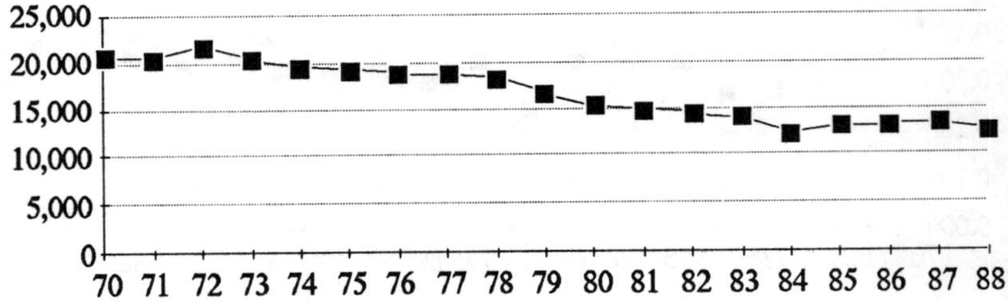

MEDLINE Records for German Documents:
Percent of All Records, by Year

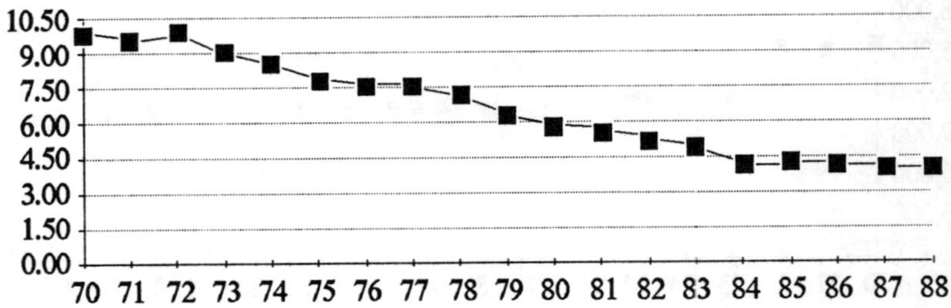

MEDLINE Records for Italian Documents:
Records by Year

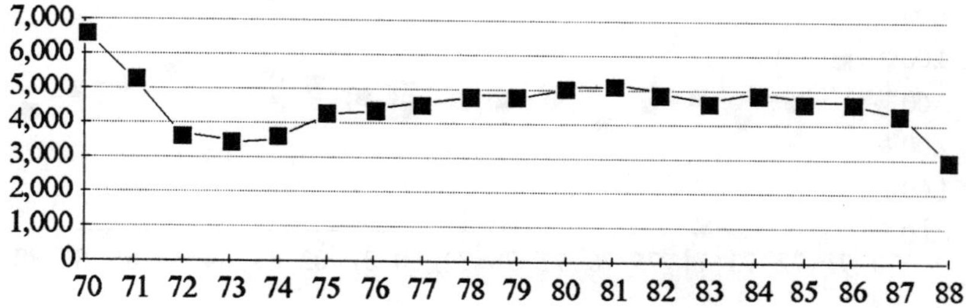

MEDLINE Records for Italian Documents:
Percent of All Records, by Year

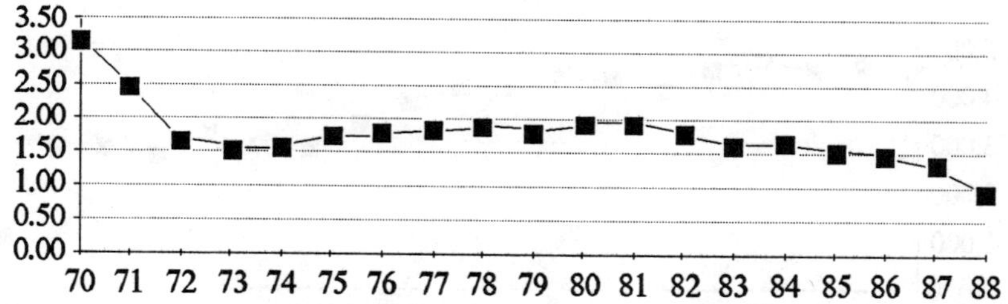

MEDLINE Records for Japanese Documents:
Records by Year

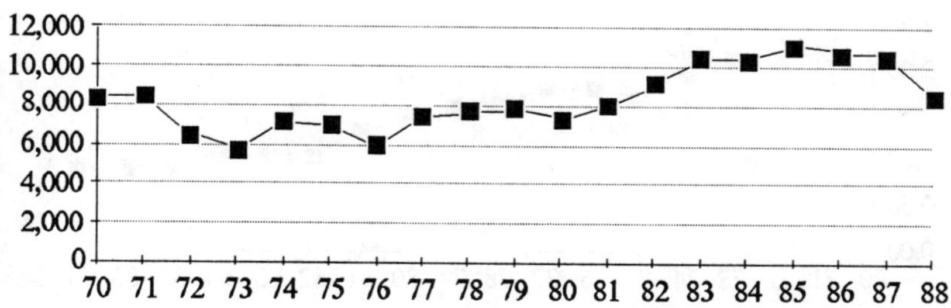

MEDLINE Records for Japanese Documents:
Percent of All Records, by Year

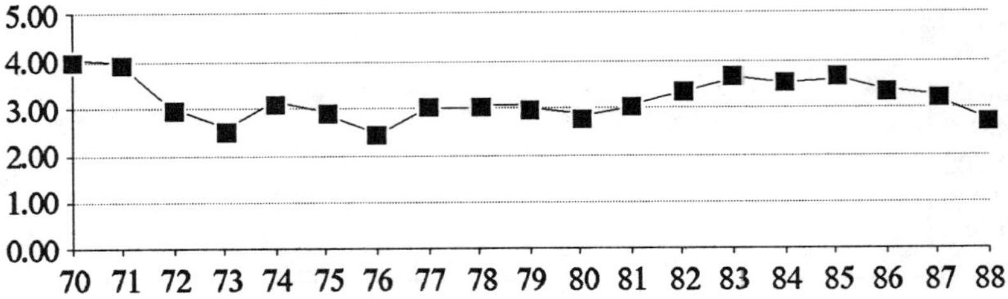

MEDLINE Records for Polish Documents:
Records by Year

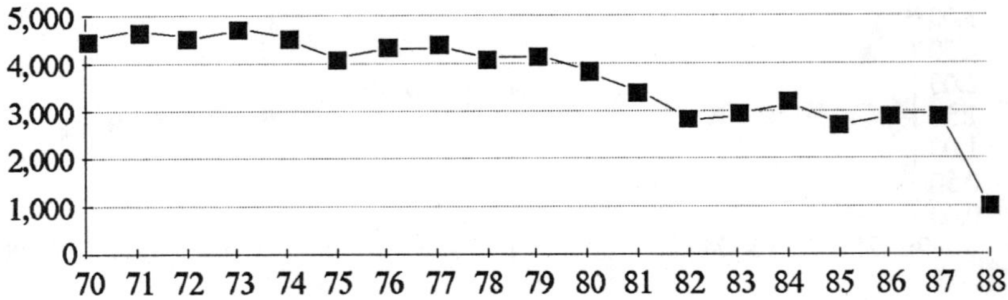

MEDLINE Records for Polish Documents:
Percent of All Records, by Year

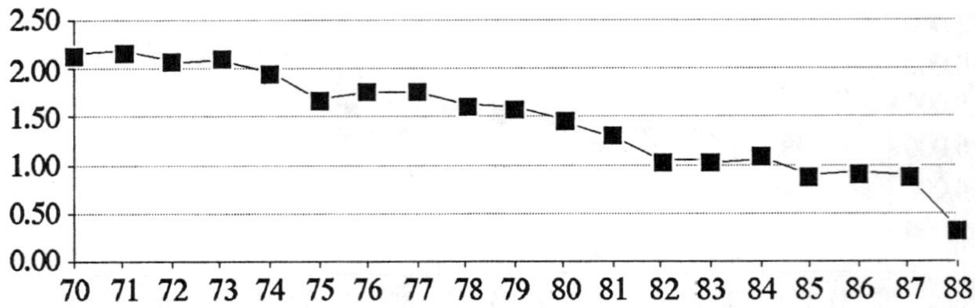

MEDLINE Records for Portuguese Documents:
Records by Year

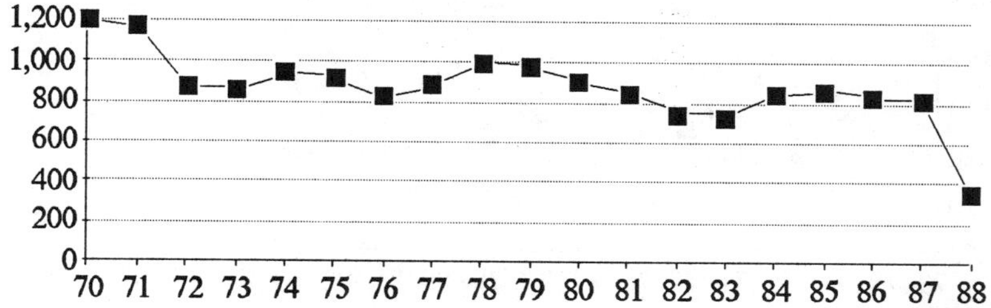

MEDLINE Records for Portuguese Documents:
Percent of All Records, by Year

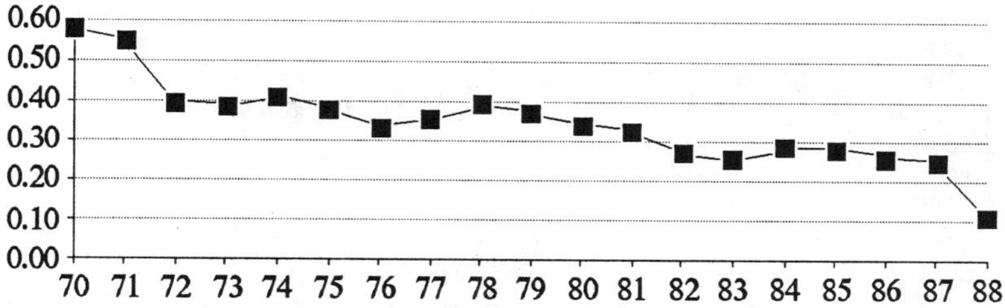

MEDLINE Records for Romanian Documents:
Records by Year

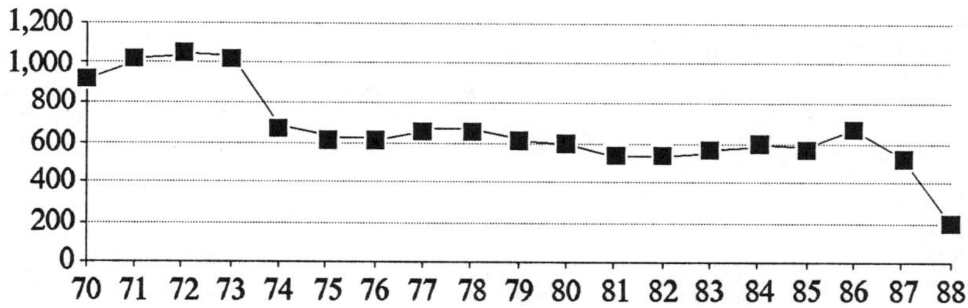

MEDLINE Records for Romanian Documents:
Percent of All Records, by Year

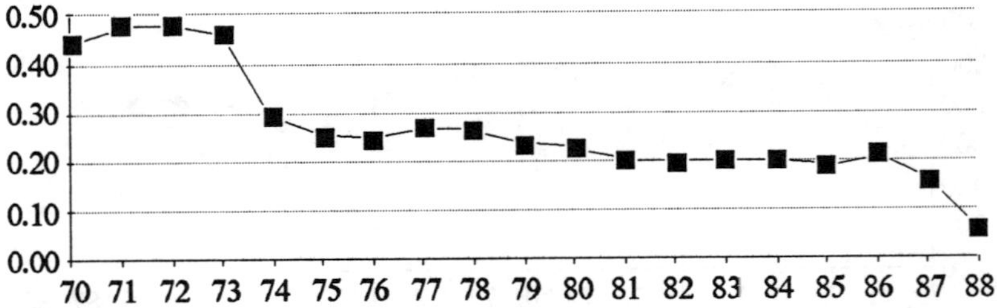

MEDLINE Records for Russian Documents:
Records by Year

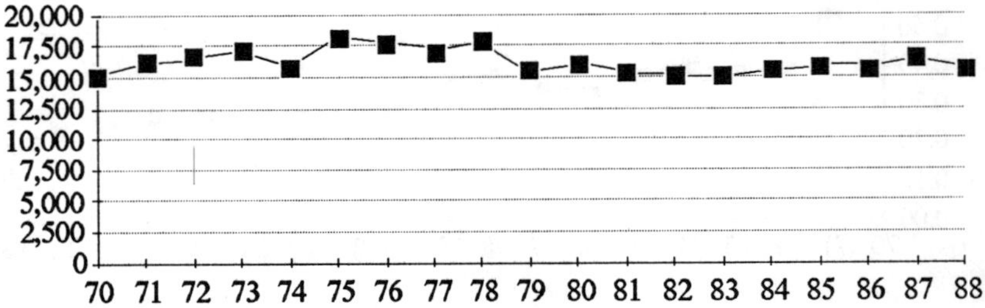

MEDLINE Records for Russian Documents:
Percent of All Records, by Year

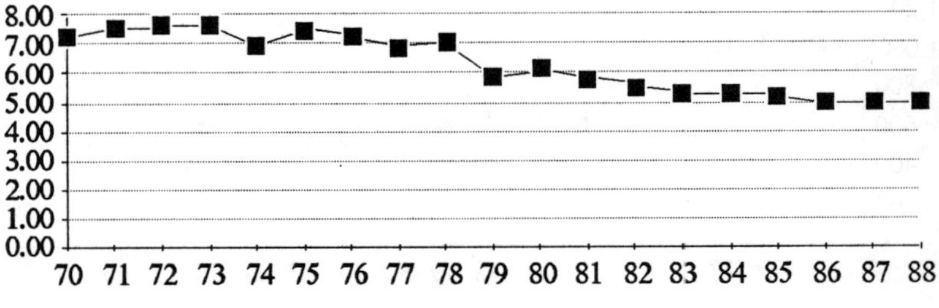

MEDLINE Records for Spanish Documents:
Records by Year

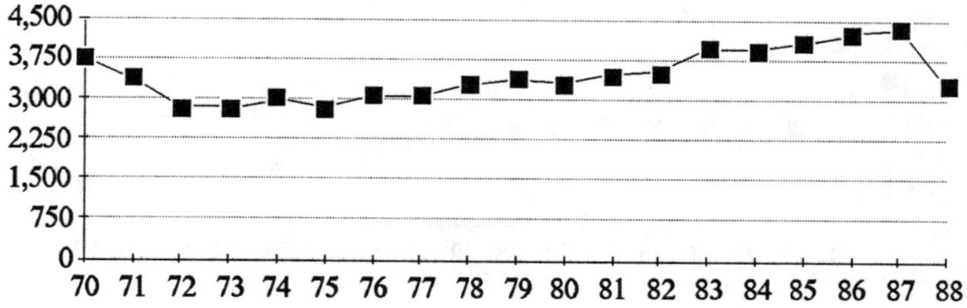

MEDLINE Records for Spanish Documents:
Percent of All Records, by Year

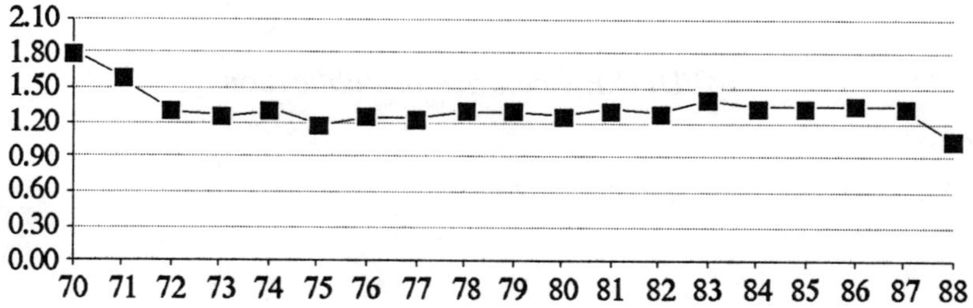

MEDLINE Records for Swedish Documents:
Records by Year

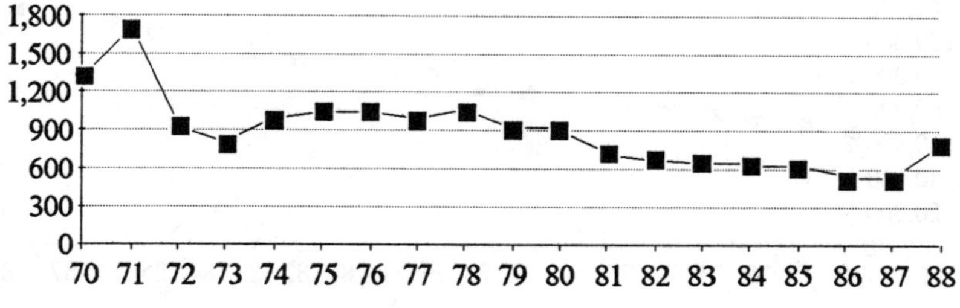

MEDLINE Records for Swedish Documents:
Percent of All Records, by Year

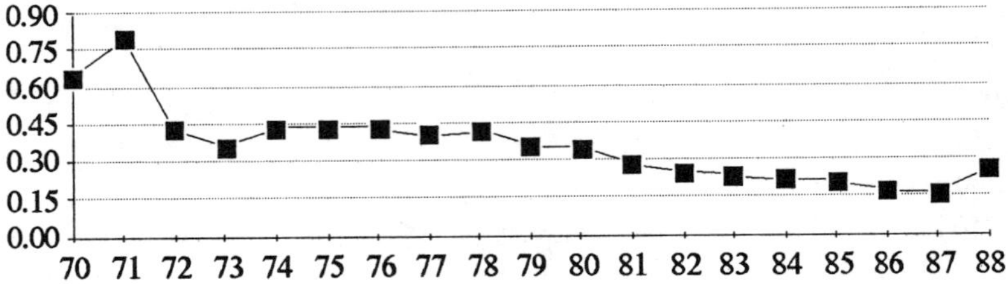

MEDLINE Countries of Publication

MEDLINE Records for Documents from the United States:
Records by Year

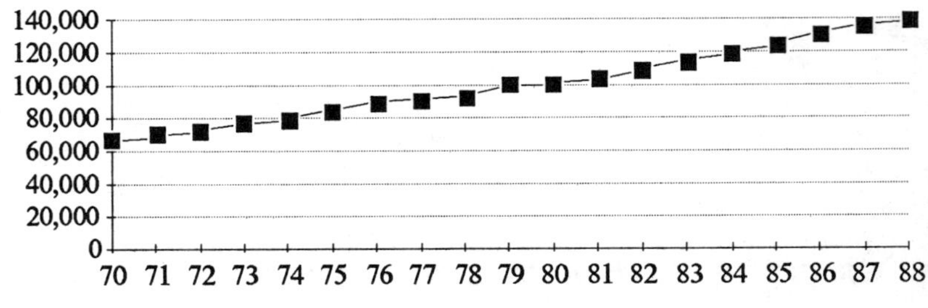

MEDLINE Records for Documents from the United States: Percent of All Records, by Year

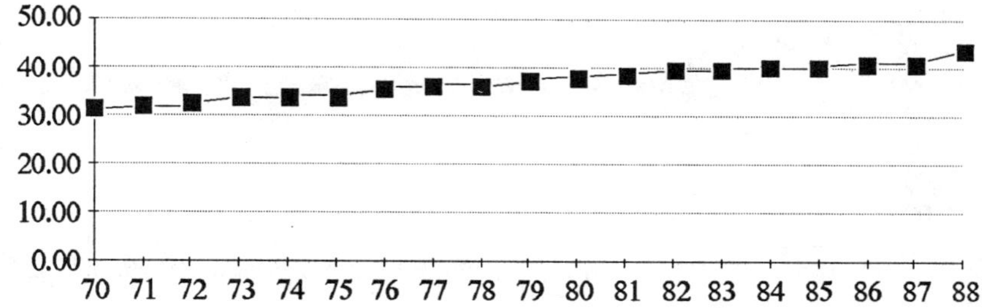

MEDLINE Records for Documents from England: Records by Year

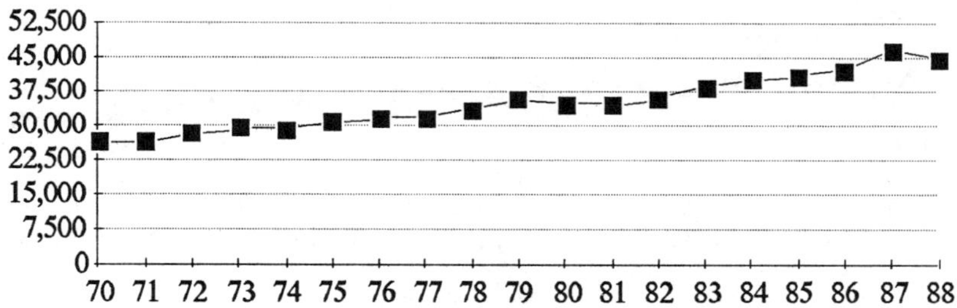

MEDLINE Records for Documents from England: Percent of All Records, by Year

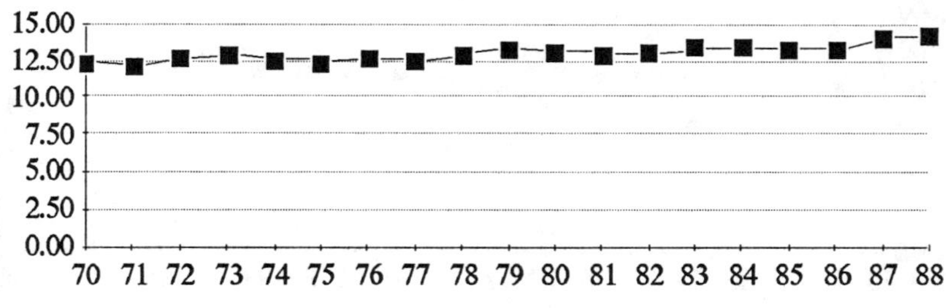

MEDLINE Records for Documents from France:
Records by Year

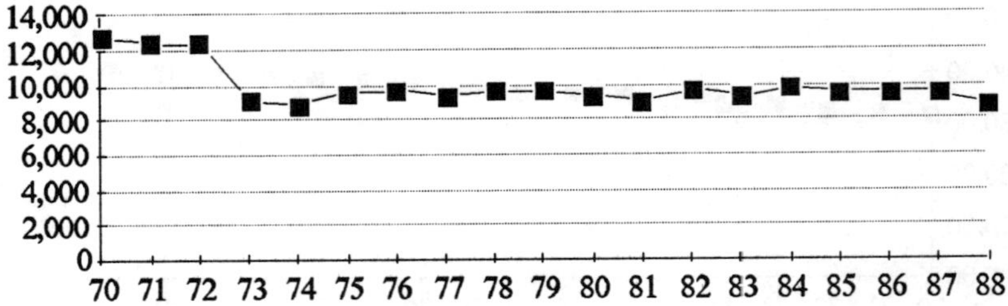

MEDLINE Records for Documents from France:
Percent of All Records, by Year

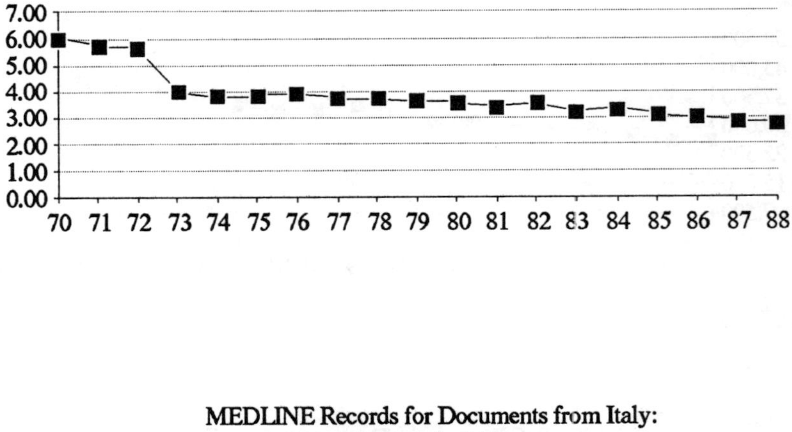

MEDLINE Records for Documents from Italy:
Records by Year

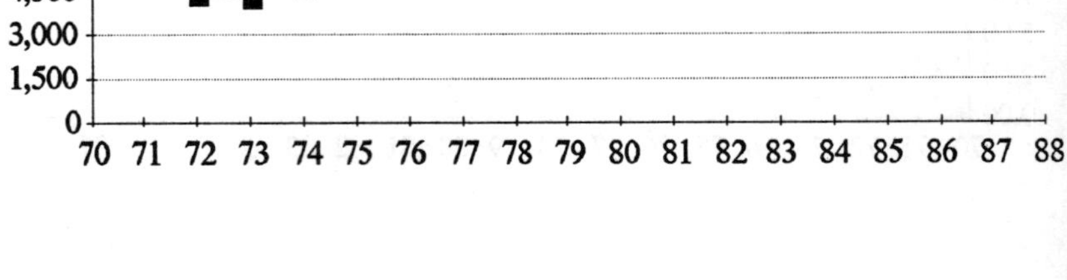

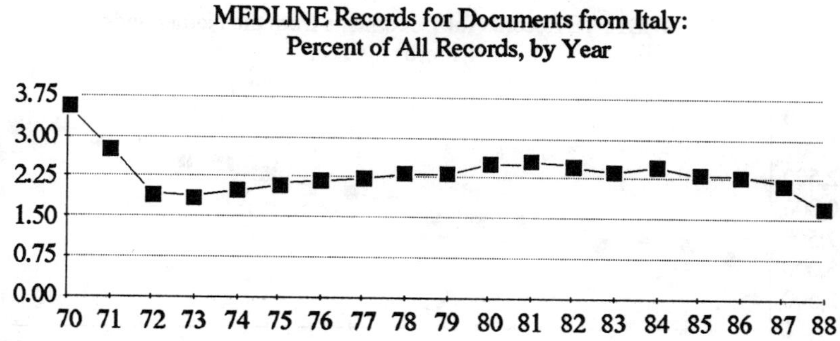
MEDLINE Records for Documents from Italy:
Percent of All Records, by Year
3.75
3.00
2.25
1.50
0.75
0.00
70 71 72 73 74 75 76 77 78 79 80 81 82 83 84 85 86 87 88

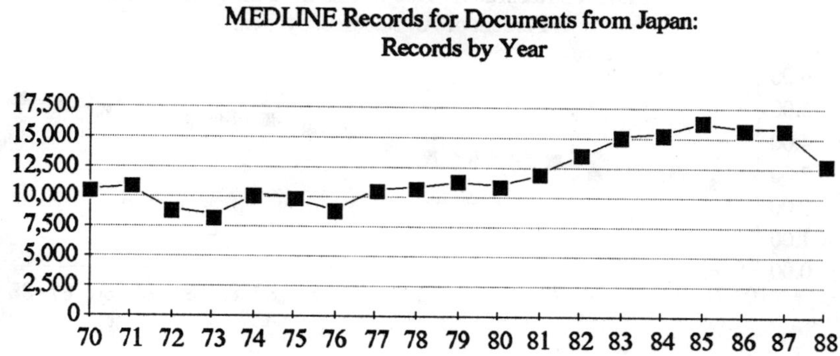
MEDLINE Records for Documents from Japan:
Records by Year
17,500
15,000
12,500
10,000
7,500
5,000
2,500
0
70 71 72 73 74 75 76 77 78 79 80 81 82 83 84 85 86 87 88

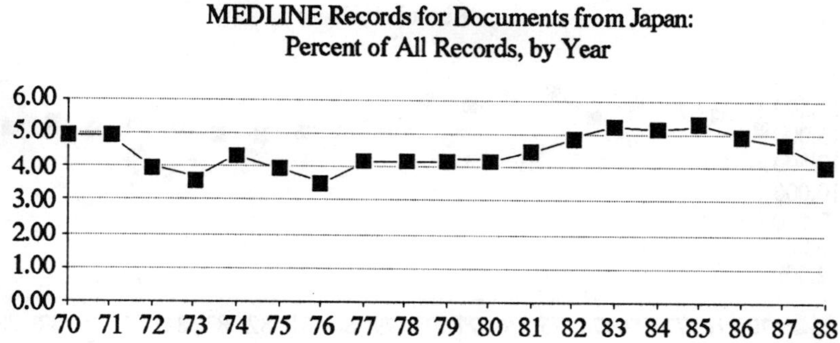
MEDLINE Records for Documents from Japan:
Percent of All Records, by Year
6.00
5.00
4.00
3.00
2.00
1.00
0.00
70 71 72 73 74 75 76 77 78 79 80 81 82 83 84 85 86 87 88

MEDLINE Records for Documents from the Netherlands:
Records by Year

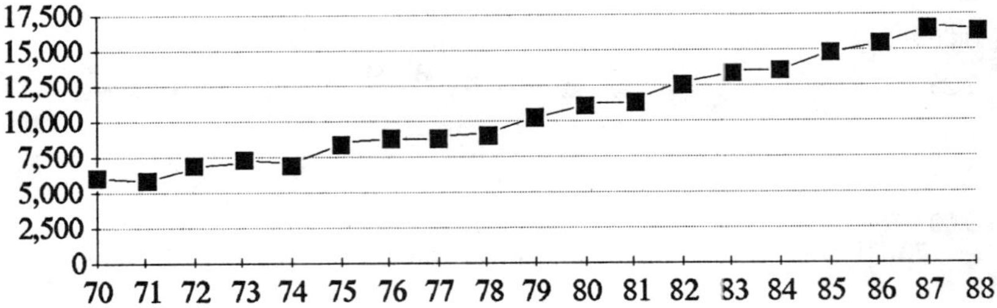

MEDLINE Records for Documents from the Netherlands:
Percent of All Records, by Year

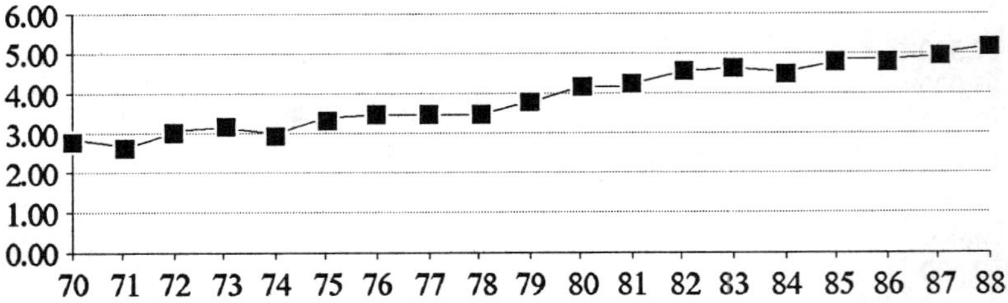

MEDLINE Records for Documents from Russia:
Records by Year

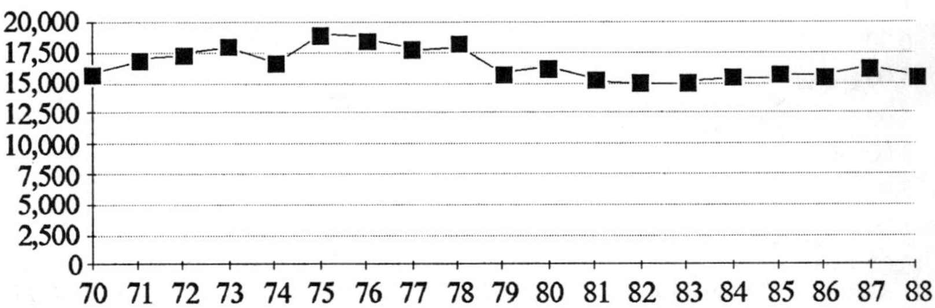

MEDLINE Records for Documents from Russia:
Percent of All Records, by Year

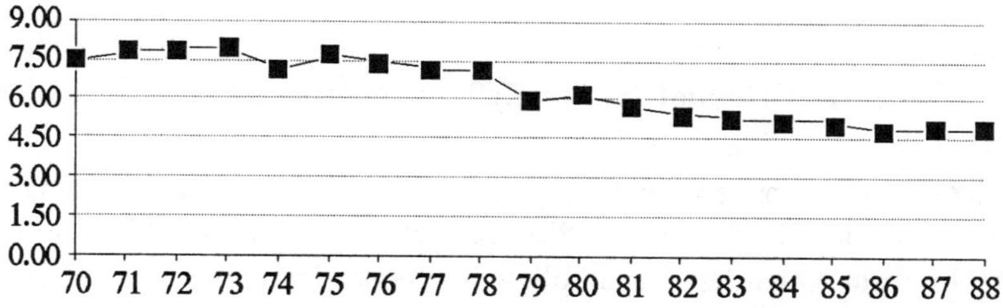

MEDLINE Records for Documents from Switzerland:
Records by Year

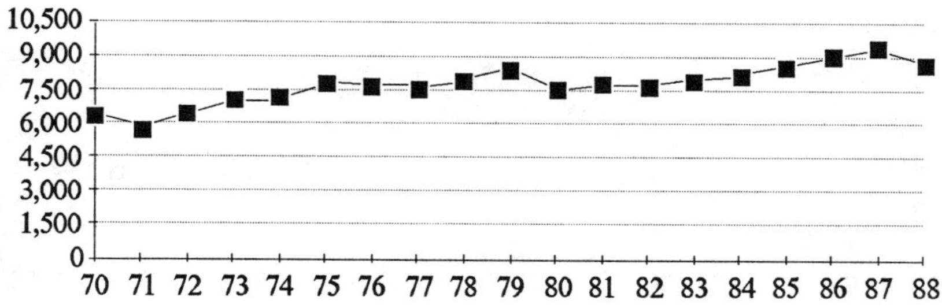

MEDLINE Records for Documents from Switzerland:
Percent of All Records, by Year

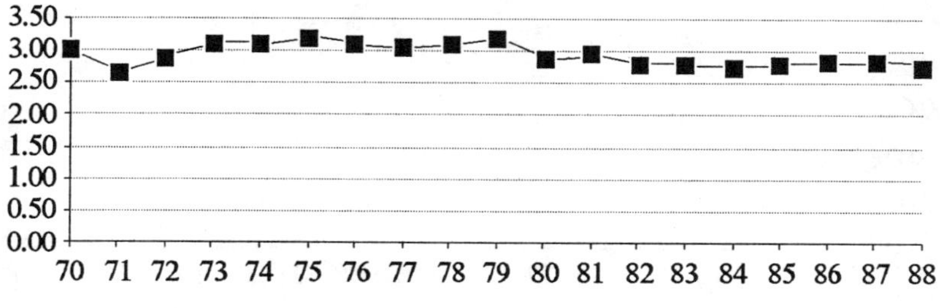

National Criminal Justice Reference Service (NCJRS)

The following charts show the trends found for languages and country of publication for the NCJRS file. Languages are presented first, led by English. Other languages follow in alphabetical order. Countries of publication follow, led by the United States, with other countries following in alphabetical order. Percentages are whole percents: That is, ".65%" indicates a sixty-five hundredths of a percent, "1.5%" indicates one and a half percent.

NCJRS English Records:
Records by Year

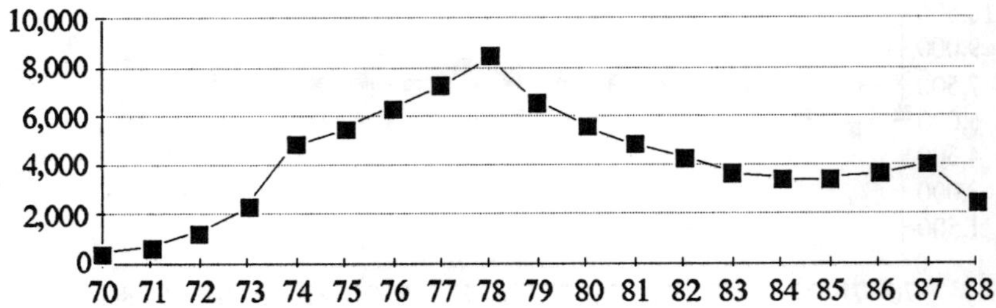

NCJRS English Records:
Percent of All Records, by Year

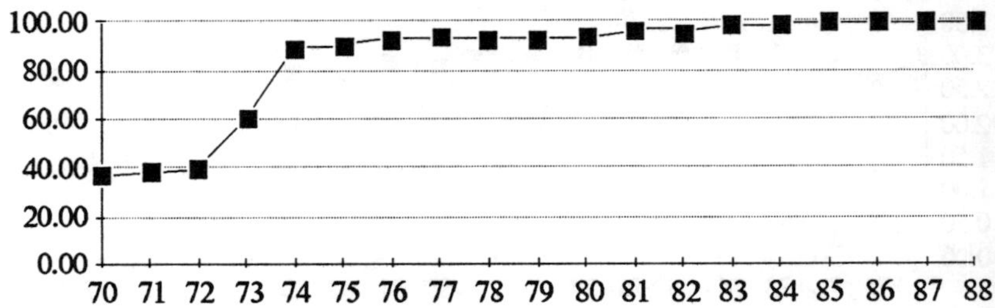

NCJRS Records for Dutch Documents:
Records by Year

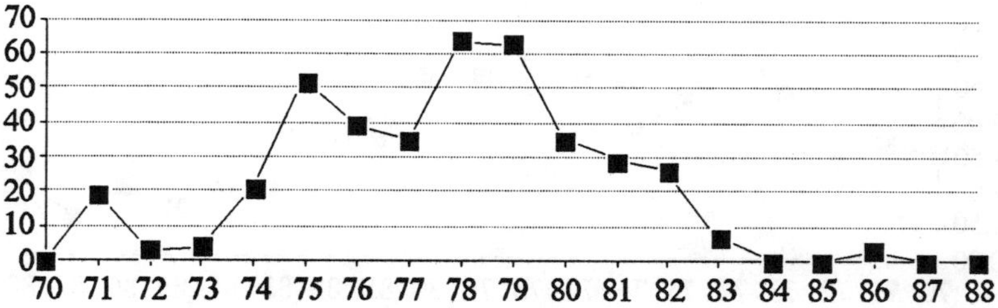

NCJRS Records for Dutch Documents:
Percent of All Records, by Year

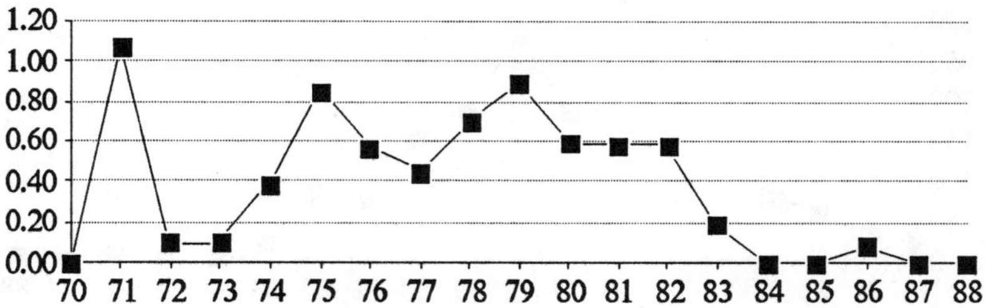

NCJRS Records for French Documents:
Records by Year

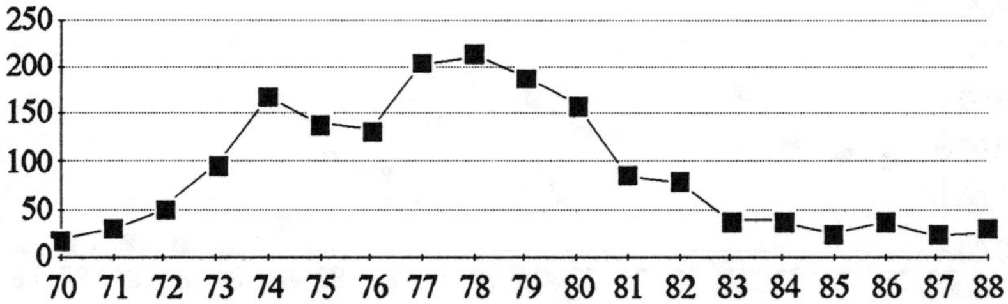

NCJRS Records for French Documents:
Percent of All Records, by Year

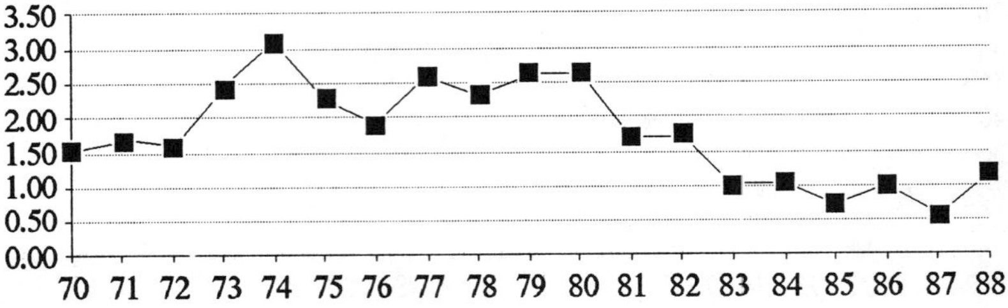

NCJRS Records for German Documents:
Records by Year

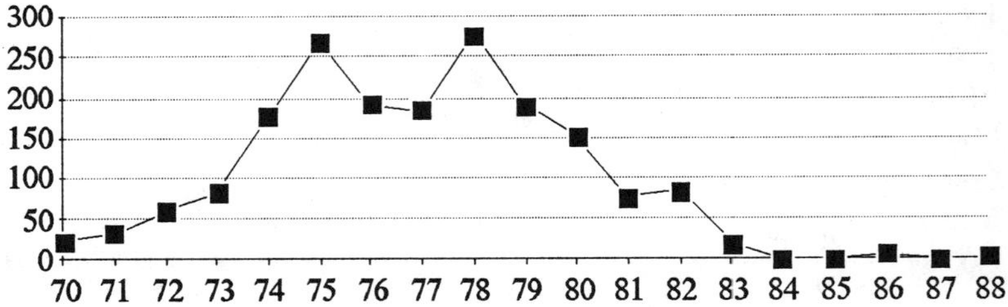

NCJRS Records for German Documents:
Percent of All Records, by Year

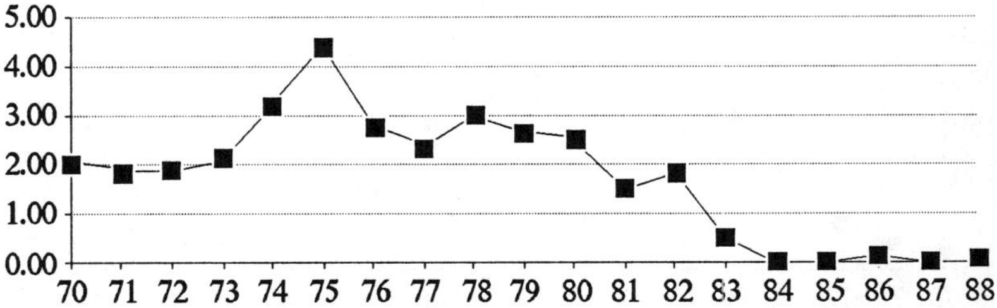

NCJRS Records for Italian Documents:
Records by Year

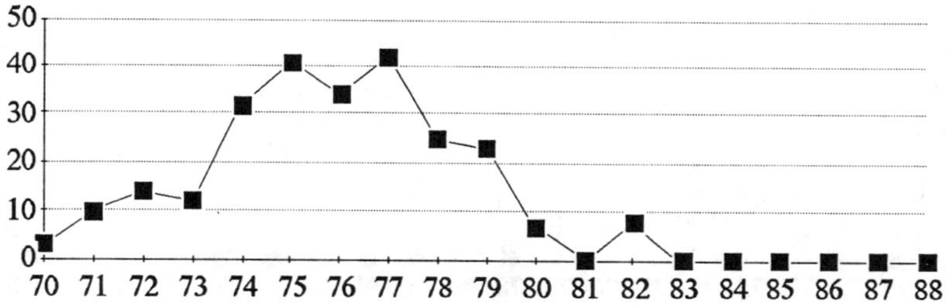

NCJRS Records for Italian Documents:
Percent of All Records, by Year

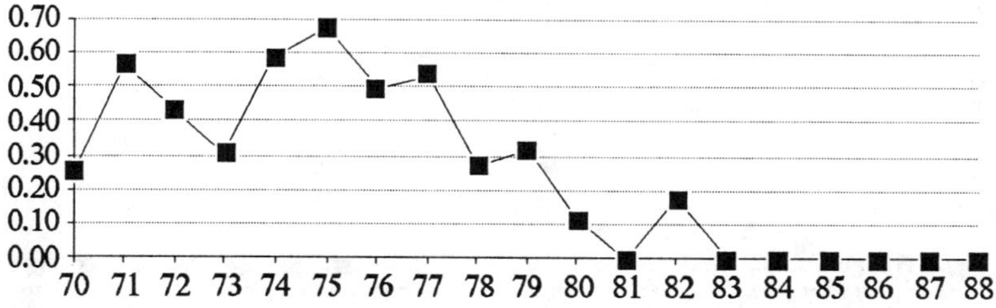

NCJRS Records for Japanese Documents:
Records by Year

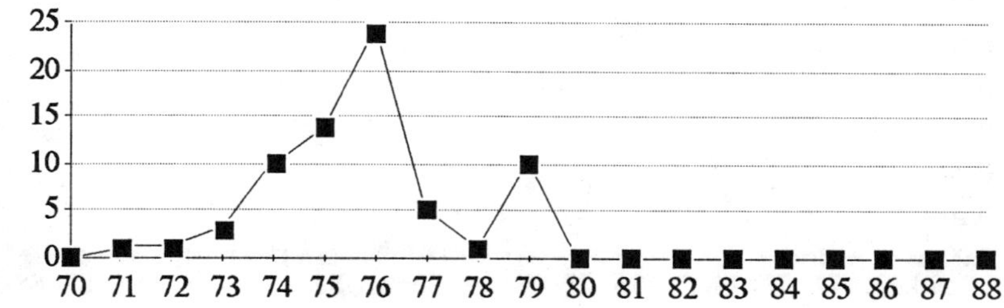

NCJRS Records for Japanese Documents:
Percent of All Records, by Year

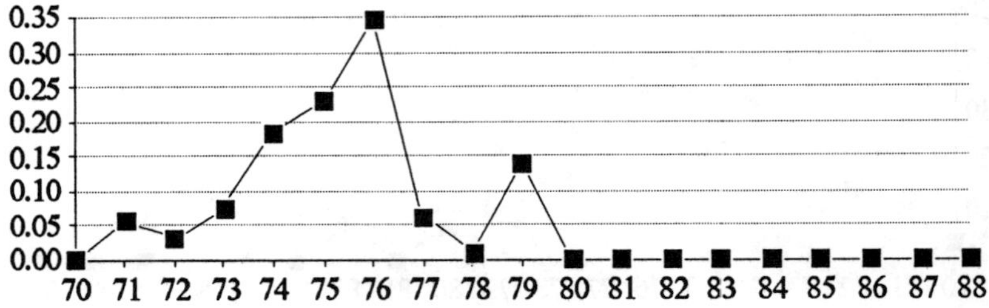

NCJRS Records for Polish Documents:
Records by Year

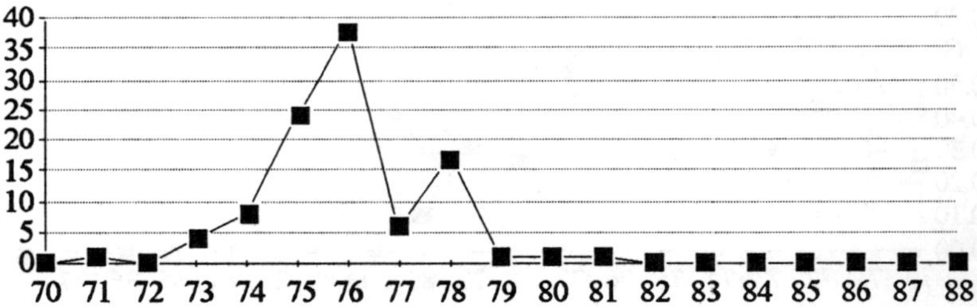

NCJRS Records for Polish Documents:
Percent of All Records, by Year

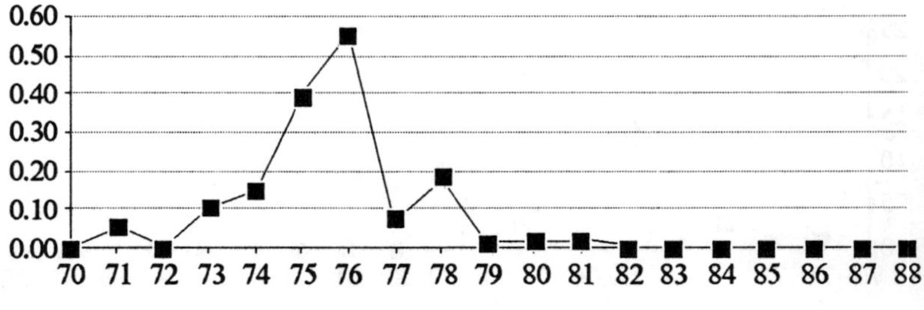

NCJRS Records for Spanish Documents:
Percent of All Records, by Year

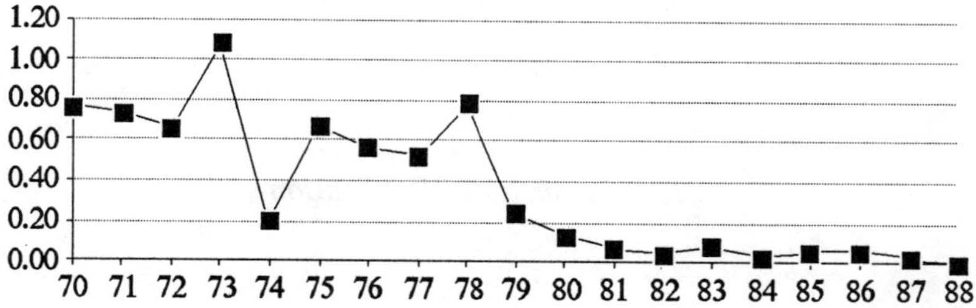

NCJRS Records for Swedish Documents:
Records by Year

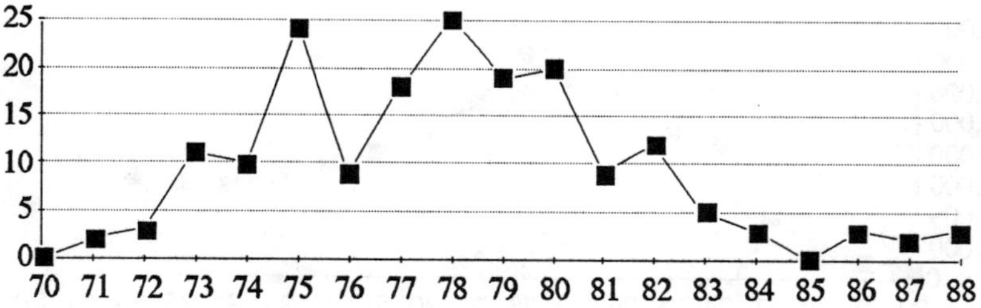

NCJRS Records for Swedish Documents:
Percent of All Records, by Year

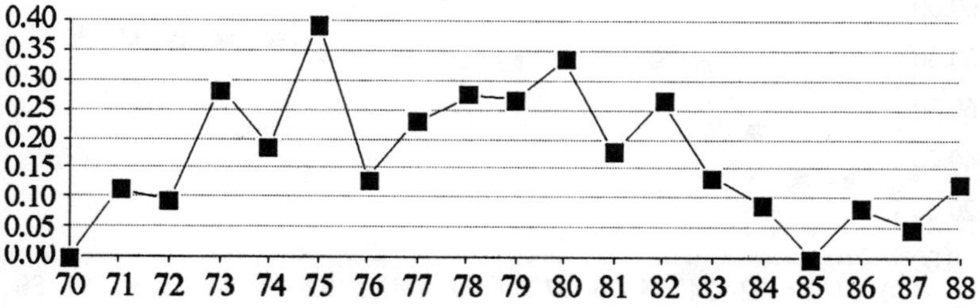

NCJRS Countries of Publication

NCJRS Records for Documents from the United States:
Records by Year

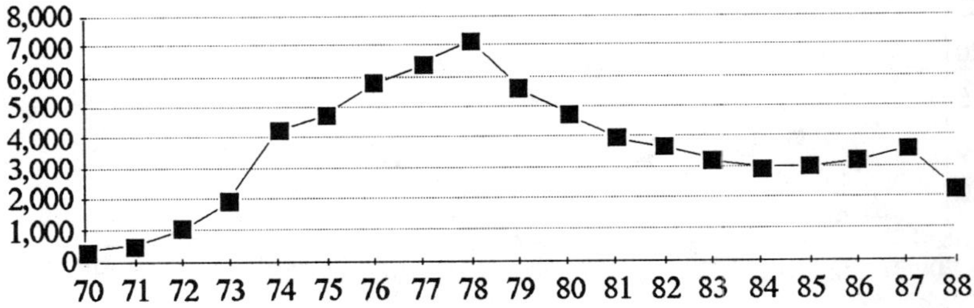

NCJRS Records for Documents from the United States:
Percent of All Records, by Year

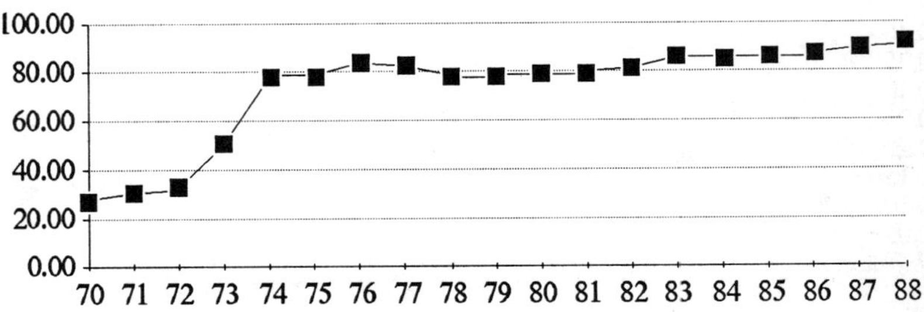

NCJRS Records for Documents from Australia:
Records by Year

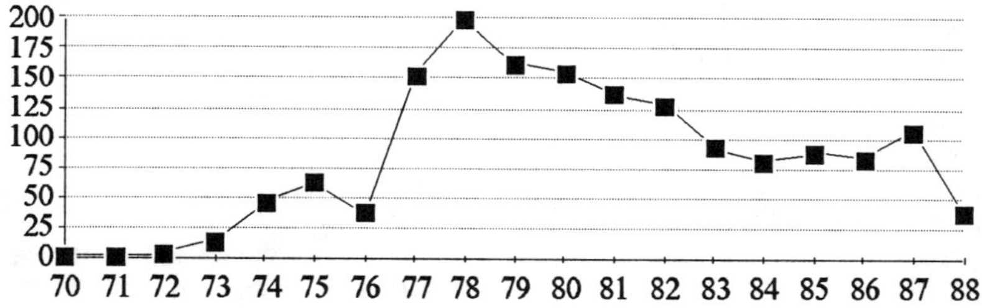

NCJRS Records for Documents from Australia:
Percent of All Records, by Year

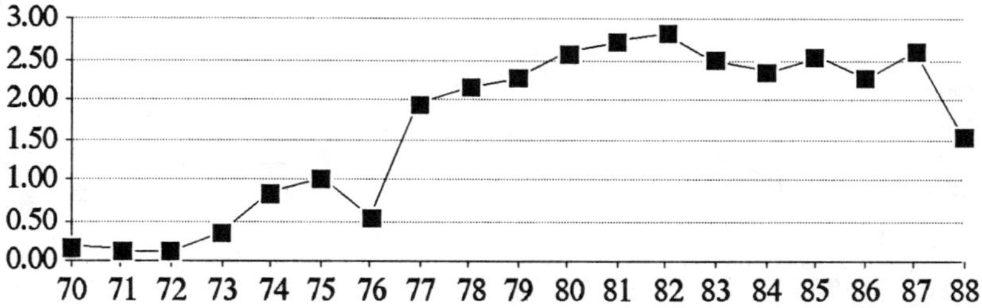

NCJRS Records for Documents from Belgium:
Records by Year

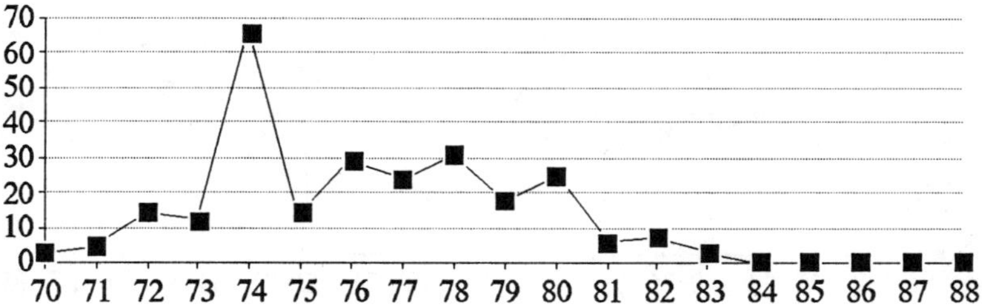

NCJRS Records for Documents from Belgium:
Percent of All Records, by Year

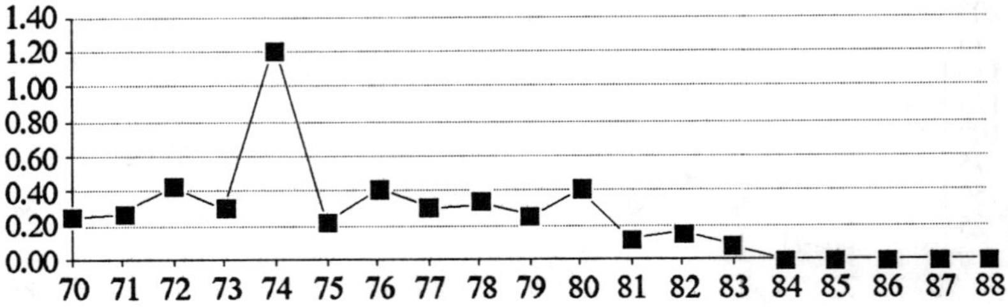

NCJRS Records for Documents from Canada:
Records by Year

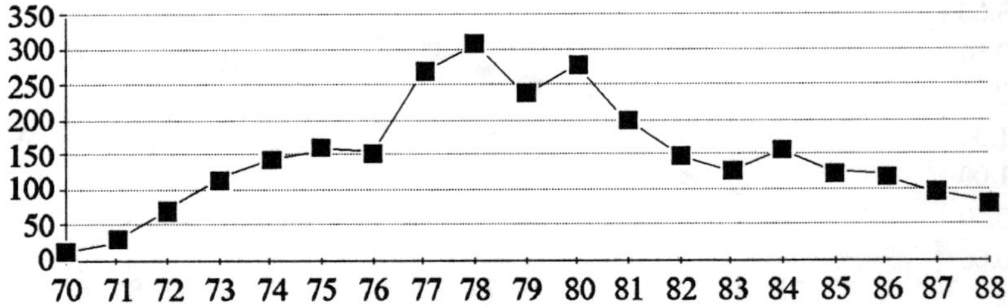

NCJRS Records for Documents from Canada:
Percent of All Records, by Year

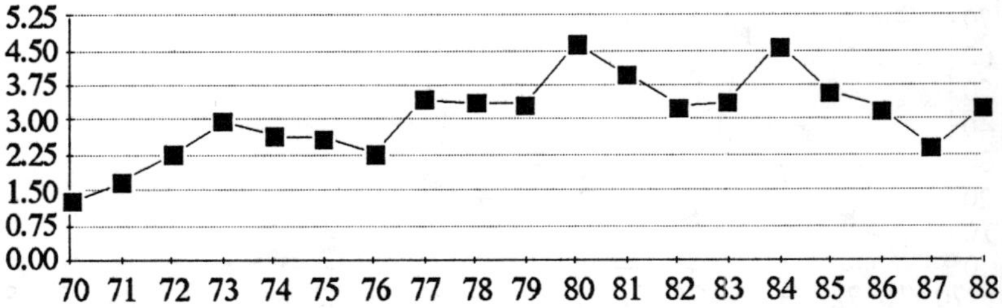

NCJRS Records for Documents from France:
Records by Year

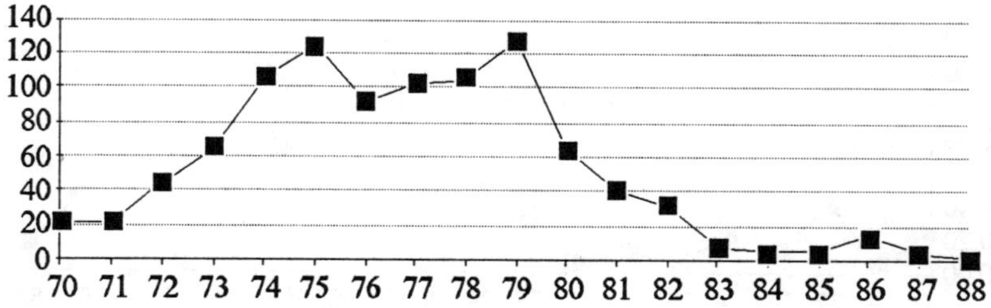

NCJRS Records for Documents from France:
Percent of All Records, by Year

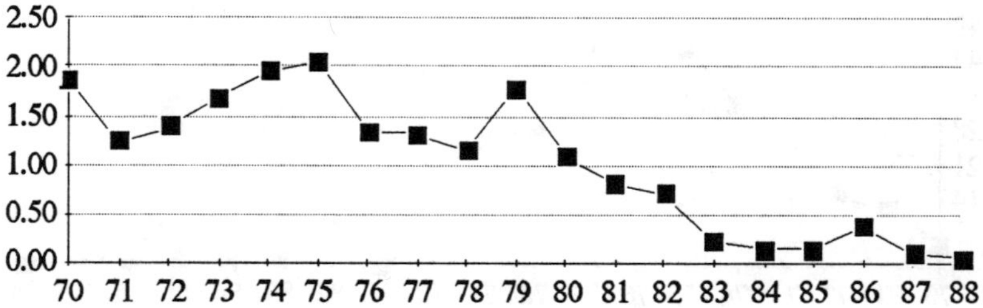

NCJRS Records for Documents from India:
Records by Year

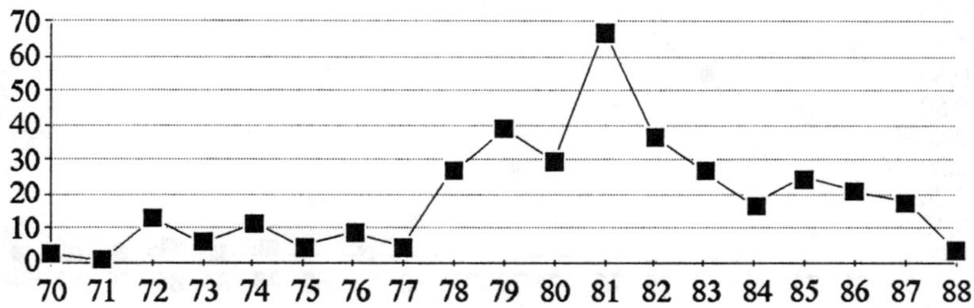

NCJRS Records for Documents from India:
Percent of All Records, by Year

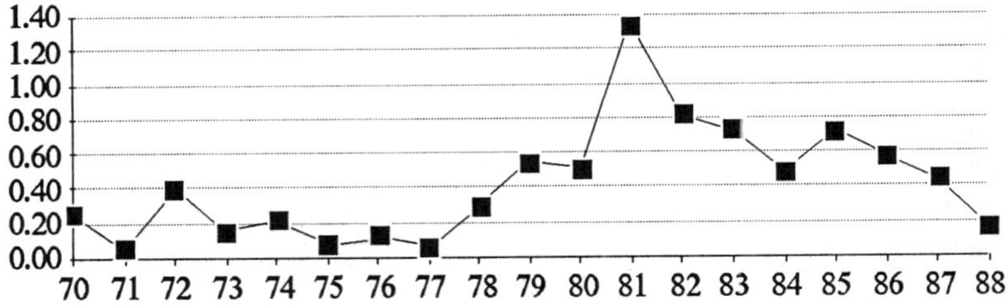

NCJRS Records for Documents from Italy:
Records by Year

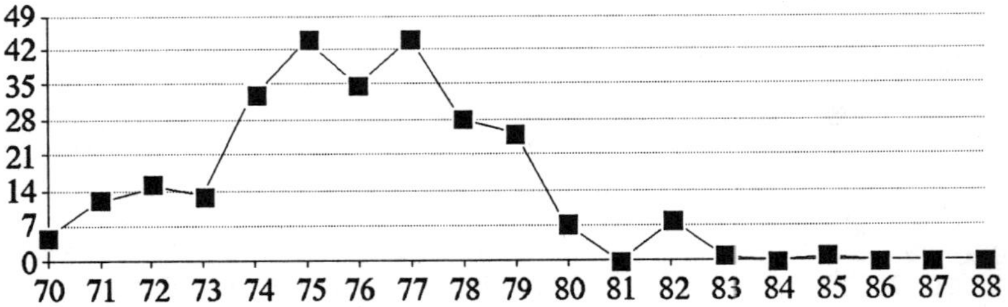

NCJRS Records for Documents from Italy:
Percent of All Records, by Year

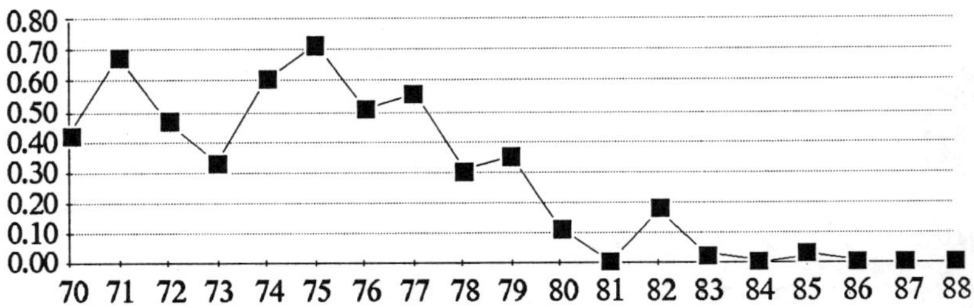

NCJRS Records for Documents from Japan:
Records by Year

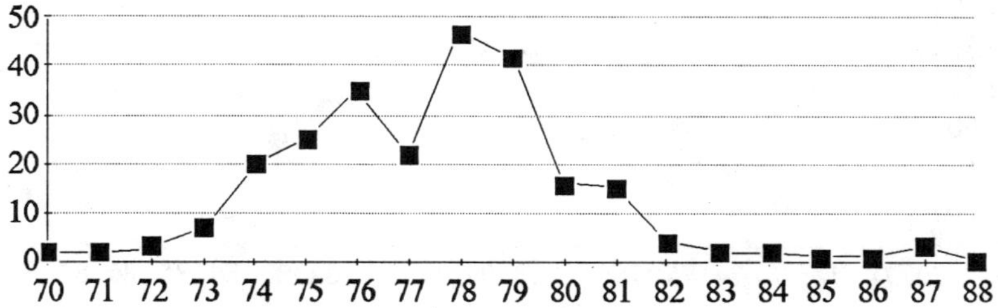

NCJRS Records for Documents from Japan:
Percent of All Records, by Year

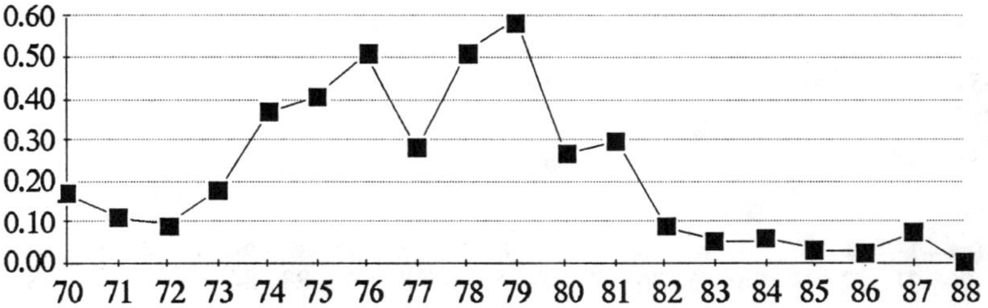

NCJRS Records for Documents from the Netherlands:
Records by Year

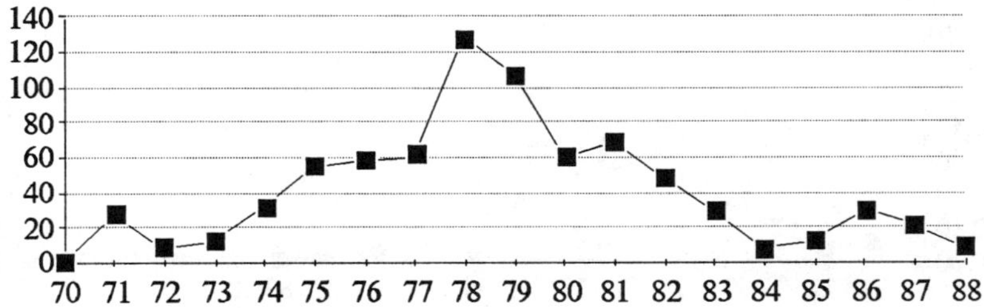

NCJRS Records for Documents from the Netherlands:
Percent of All Records, by Year

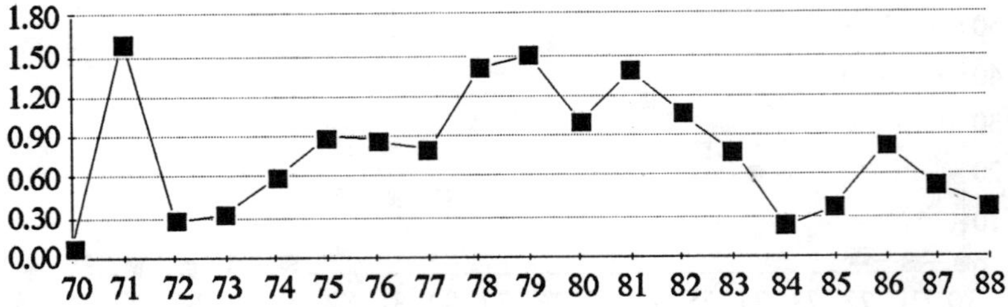

NCJRS Records for Documents from Poland:
Records by Year

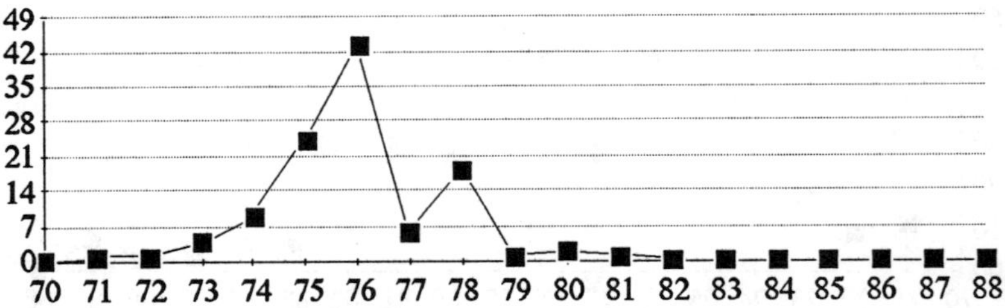

NCJRS Records for Documents from Poland:
Percent of All Records, by Year

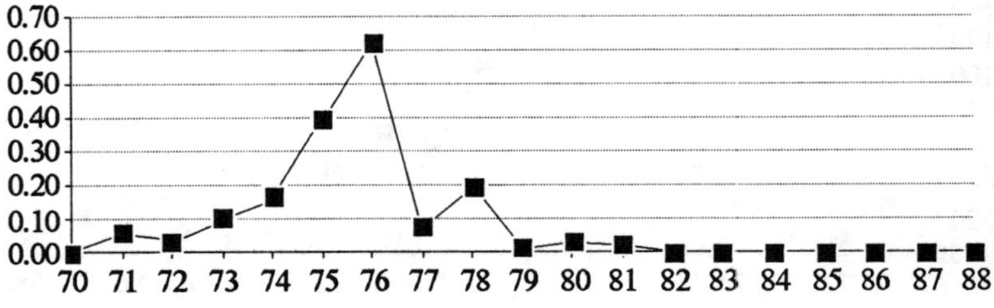

NCJRS Records for Documents from Sweden:
Records by Year

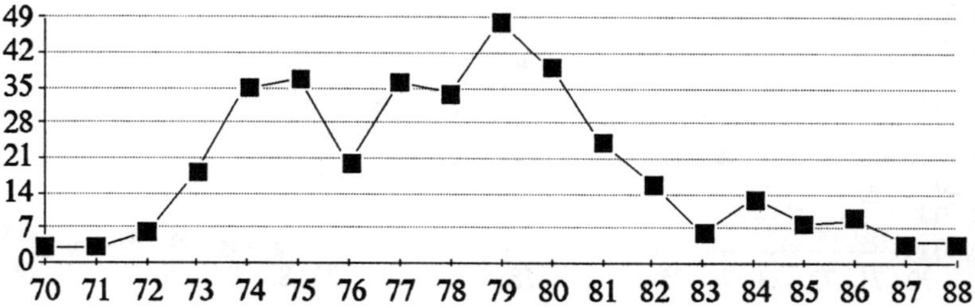

NCJRS Records for Documents from Sweden:
Percent of All Records, by Year

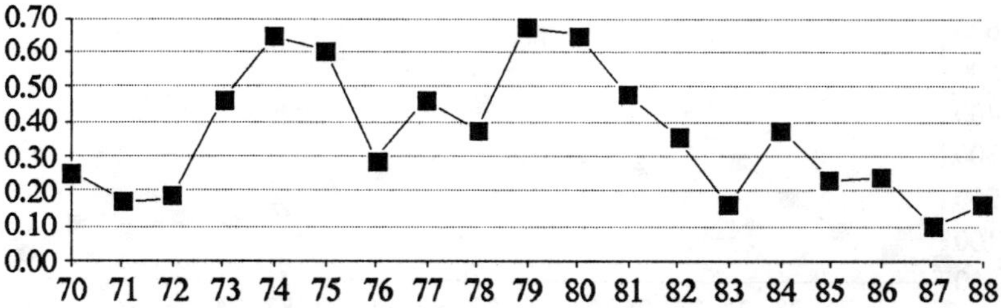

NCJRS Records for Documents from Switzerland:
Records by Year

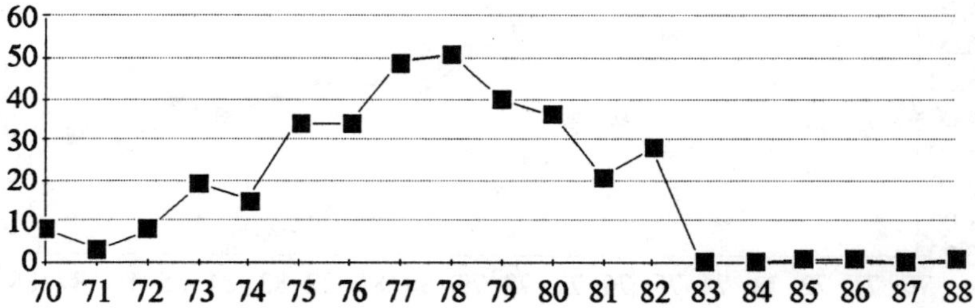

NCJRS Records for Documents from Switzerland:
Percent of All Records, by Year

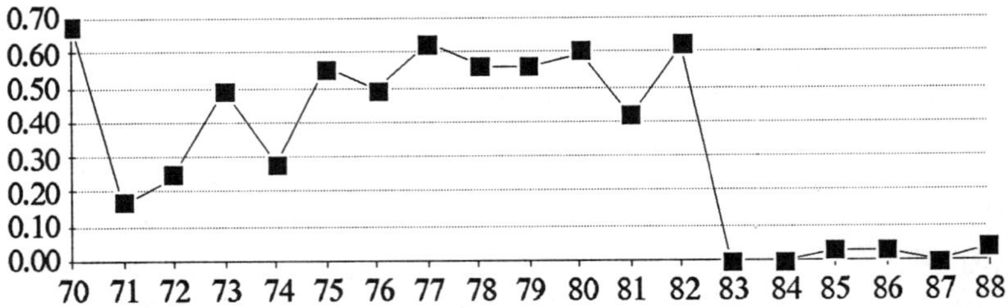

NCJRS Records for Documents from the United Kingdom:
Records by Year

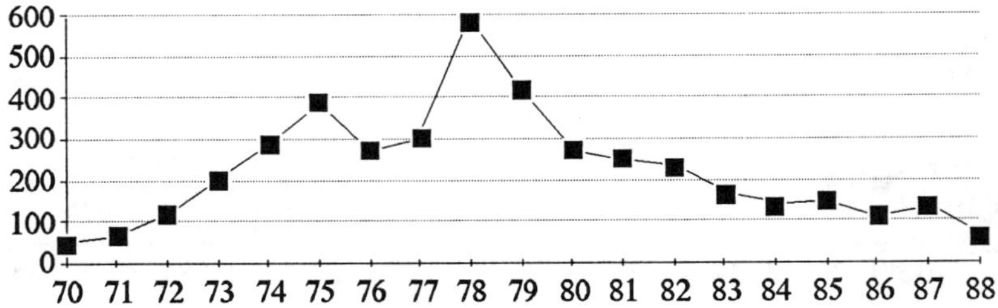

NCJRS Records for Documents from the United Kingdom:
Percent of All Records, by Year

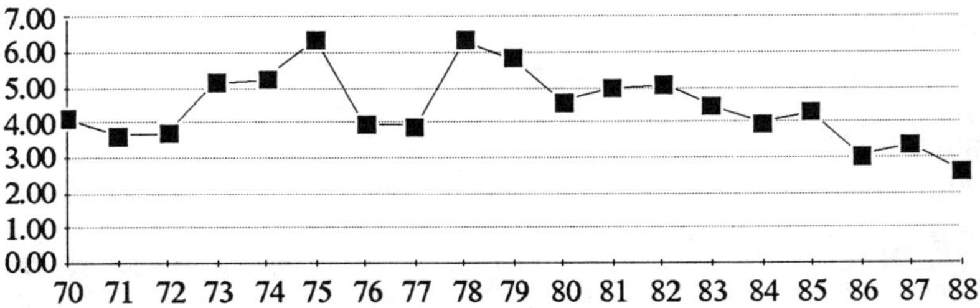

NCJRS Records for Documents from the "United Nations":
Records by Year

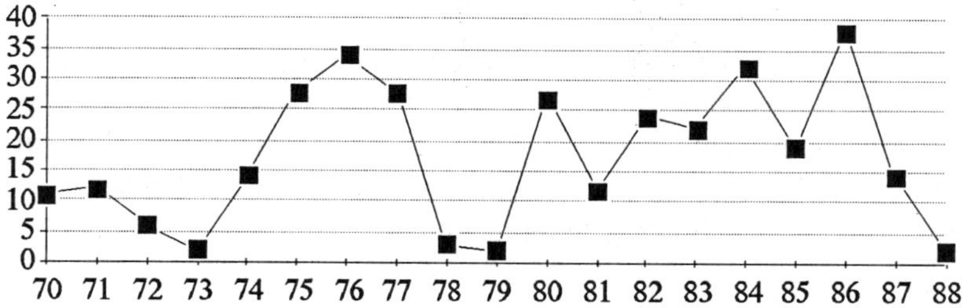

NCJRS Records for Documents from the "United Nations":
Percent of All Records, by Year

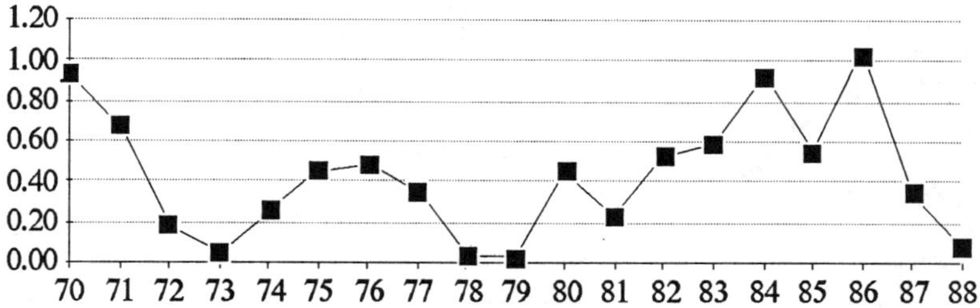

Oceanic Abstracts Records

The following charts show the trends found for primary languages and summary languages for the Oceanic Abstracts file. Primary languages are presented first, led by English. Other languages follow in alphabetical order. Summary languages follow, led by English, with other languages following in alphabetical order.

Percentages are whole percents: that is, ".65%" indicates a sixty-five hundredths of a percent, "1.5%" indicates one and a half percent.

Oceanic Abstracts Records for English Documents:
Records by Year

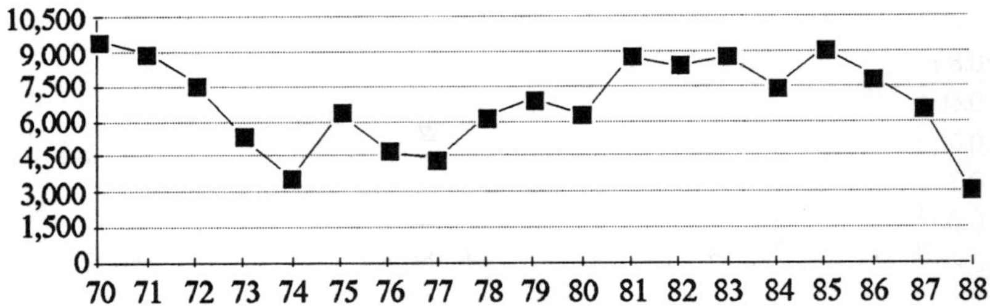

Oceanic Abstracts Records for English Documents:
Percent of All Records, by Year

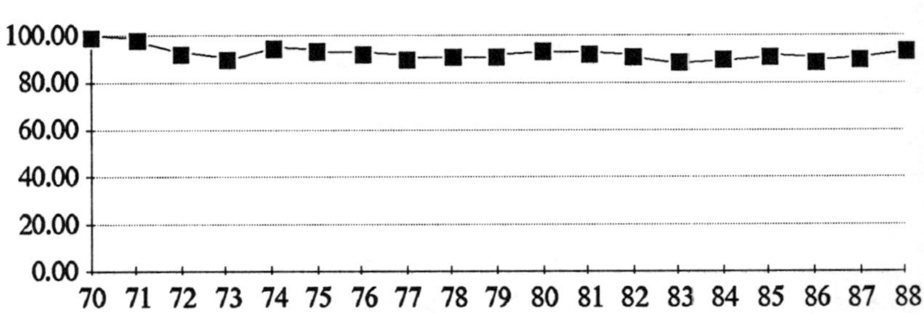

Oceanic Abstracts Records for Chinese Documents:
Records by Year

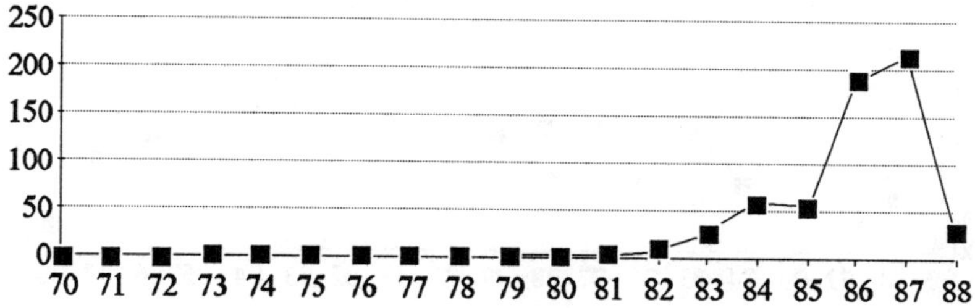

Oceanic Abstracts Records for Chinese Documents:
Percent of All Records, by Year

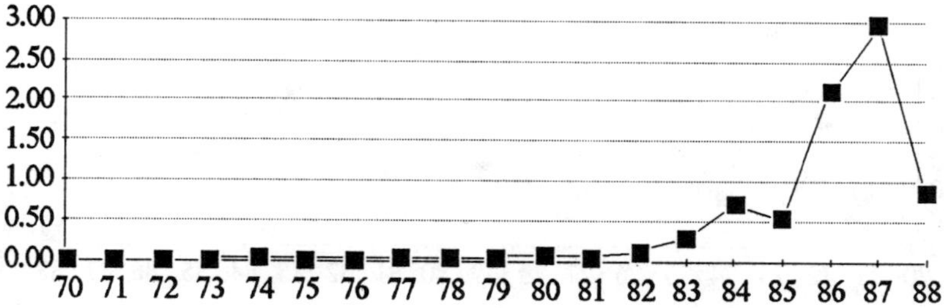

Oceanic Abstracts Records for French Documents:
Records by Year

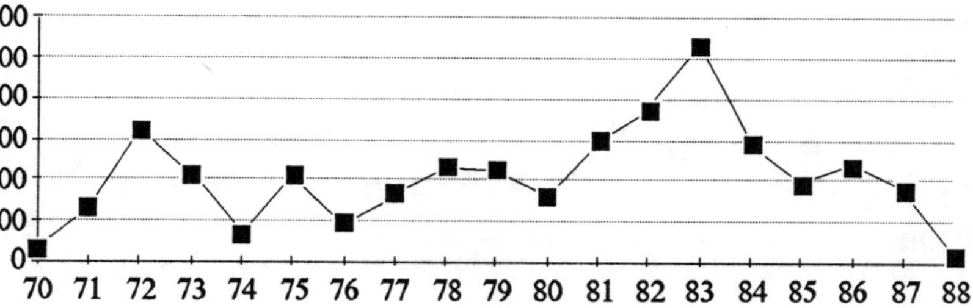

Oceanic Abstracts Records for French Documents:
Percent of All Records, by Year

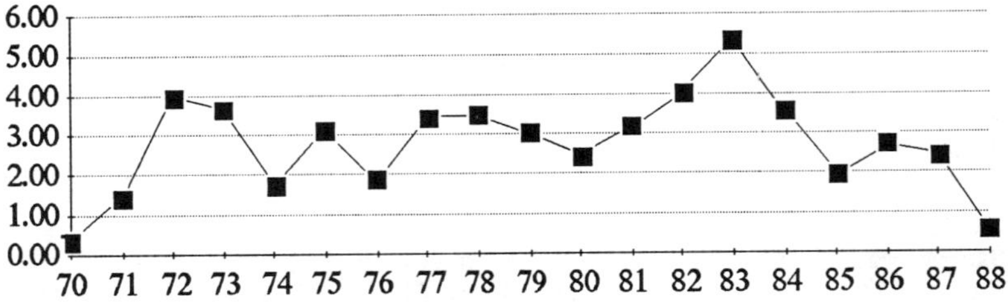

Oceanic Abstracts Records for German Documents:
Records by Year

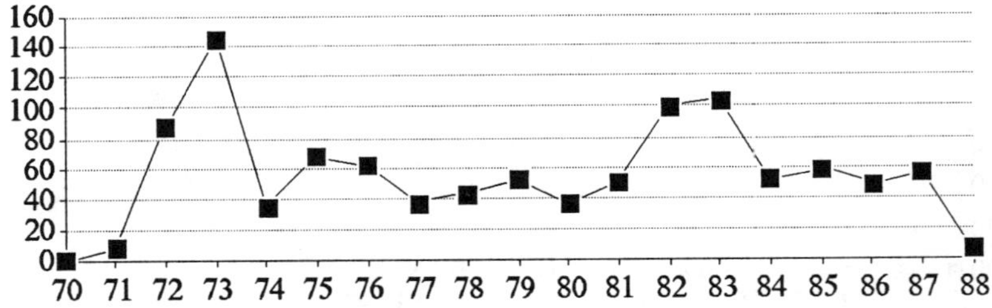

Oceanic Abstracts Records for German Documents:
Percent of All Records, by Year

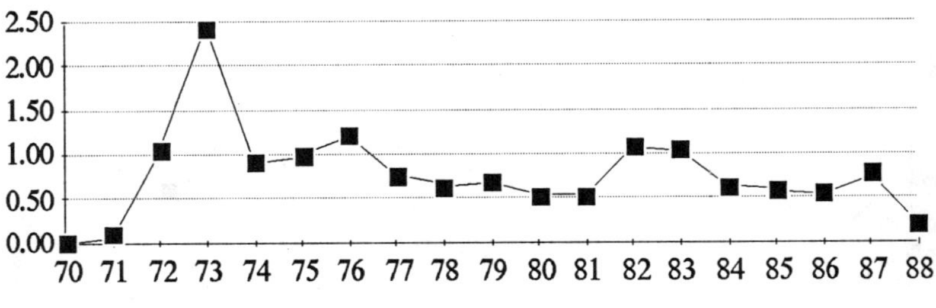

Oceanic Abstracts Records for Italian Documents:
Records by Year

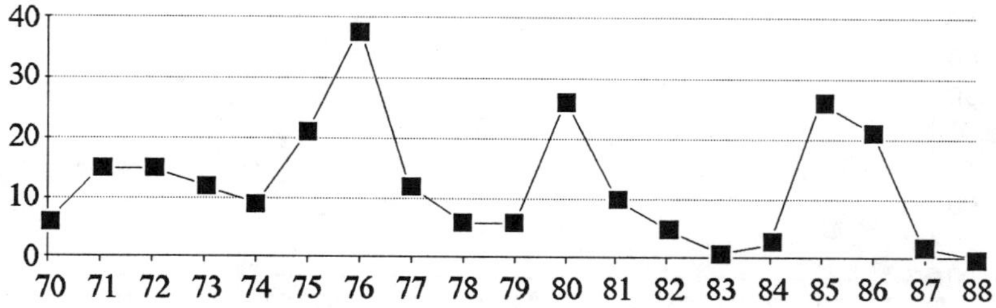

Oceanic Abstracts Records for Italian Documents:
Percent of All Records, by Year

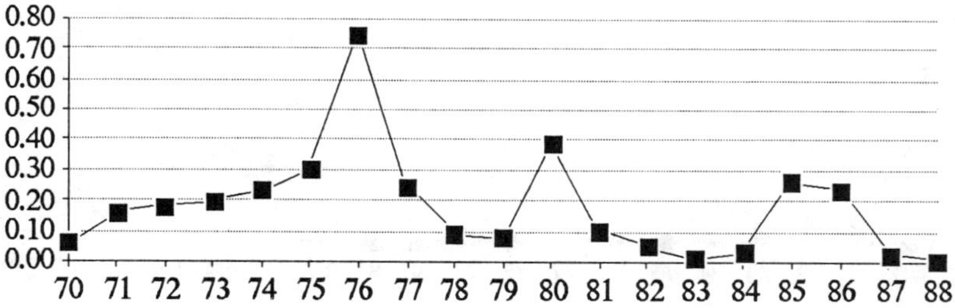

Oceanic Abstracts Records for Japanese Documents:
Records by Year

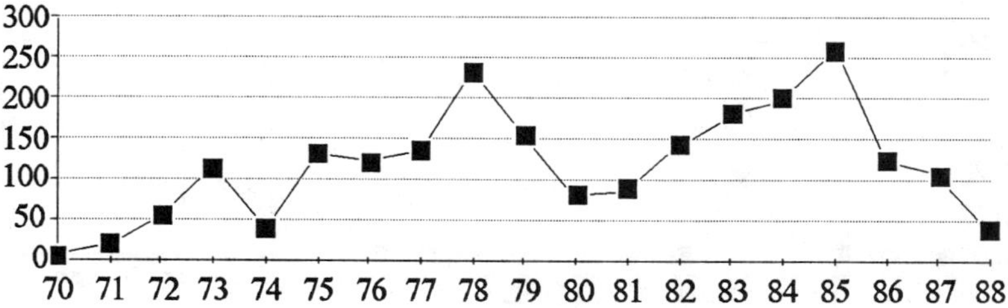

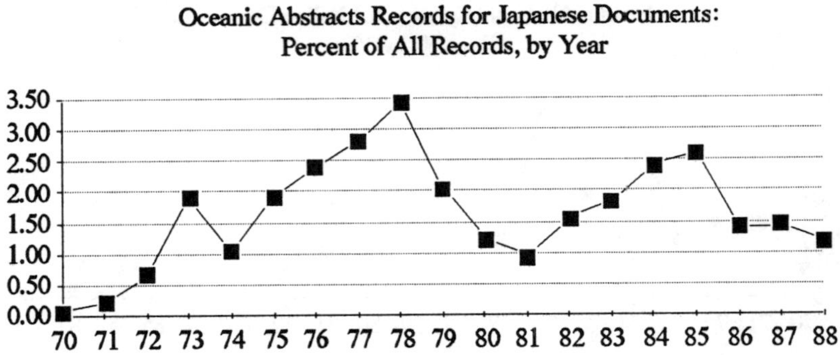
Oceanic Abstracts Records for Japanese Documents:
Percent of All Records, by Year
3.50
3.00
2.50
2.00
1.50
1.00
0.50
0.00
70 71 72 73 74 75 76 77 78 79 80 81 82 83 84 85 86 87 88

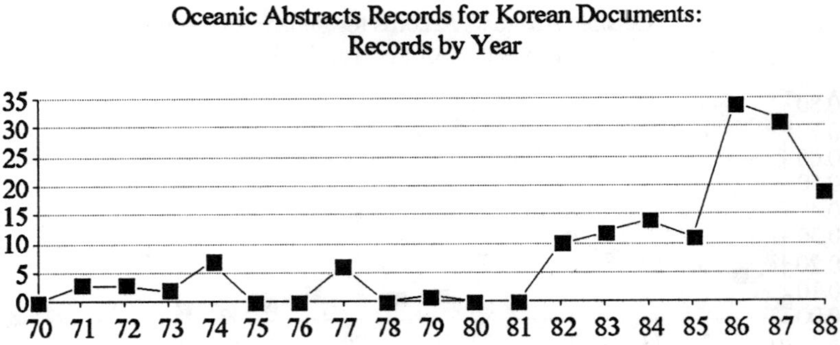
Oceanic Abstracts Records for Korean Documents:
Records by Year
35
30
25
20
15
10
5
0
70 71 72 73 74 75 76 77 78 79 80 81 82 83 84 85 86 87 88

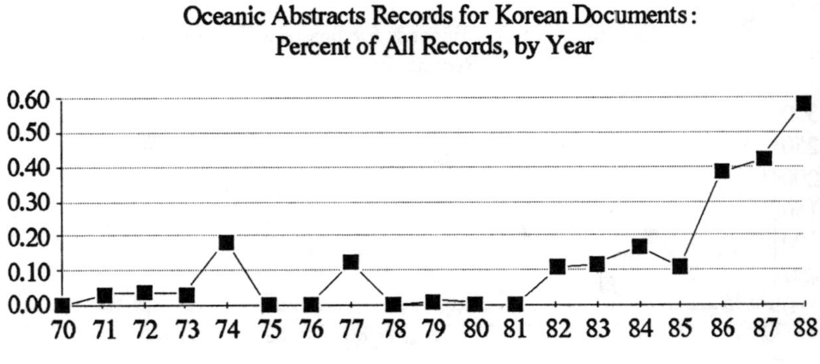
Oceanic Abstracts Records for Korean Documents:
Percent of All Records, by Year
0.60
0.50
0.40
0.30
0.20
0.10
0.00
70 71 72 73 74 75 76 77 78 79 80 81 82 83 84 85 86 87 88

Oceanic Abstracts Records for Norwegian Documents:
Records by Year

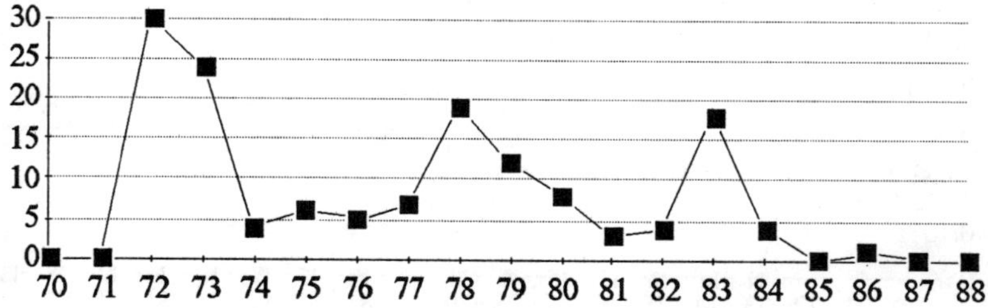

Oceanic Abstracts Records for Norwegian Documents:
Percent of All Records, by Year

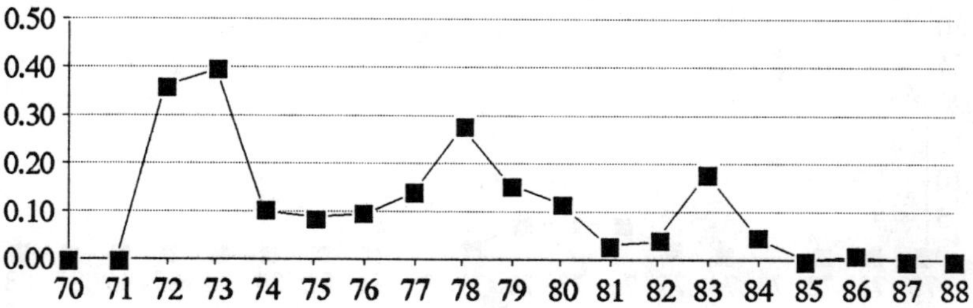

Oceanic Abstracts Records for "Persian" Documents:
Records by Year

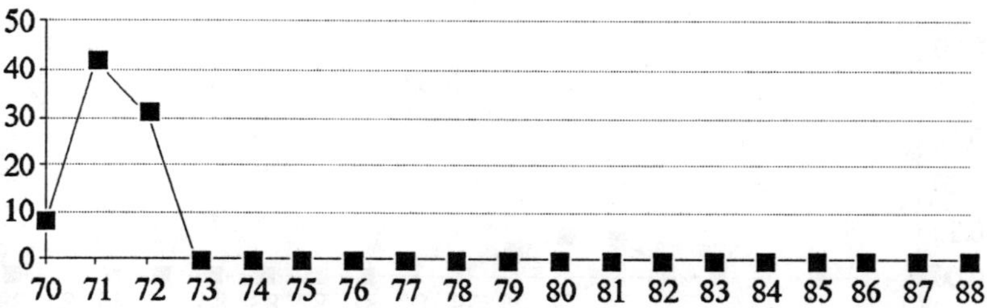

Oceanic Abstracts Records for "Persian" Documents:
Percent of All Records, by Year

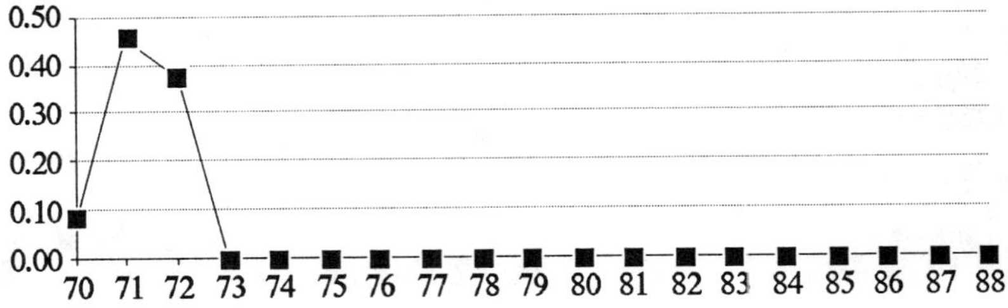

Oceanic Abstracts Records for Polish Documents:
Records by Year

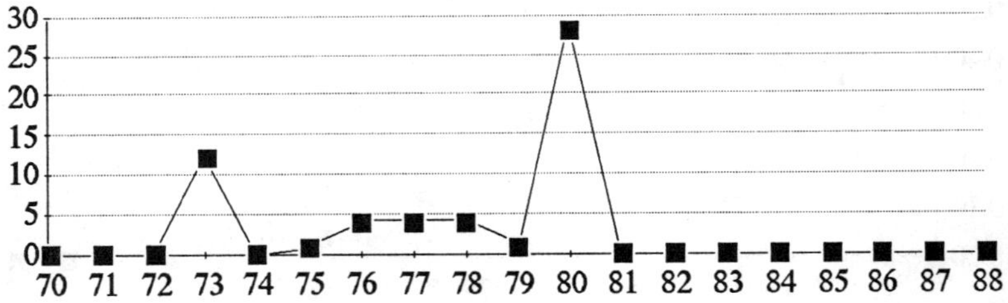

Oceanic Abstracts Records for Polish Documents:
Percent of All Records, by Year

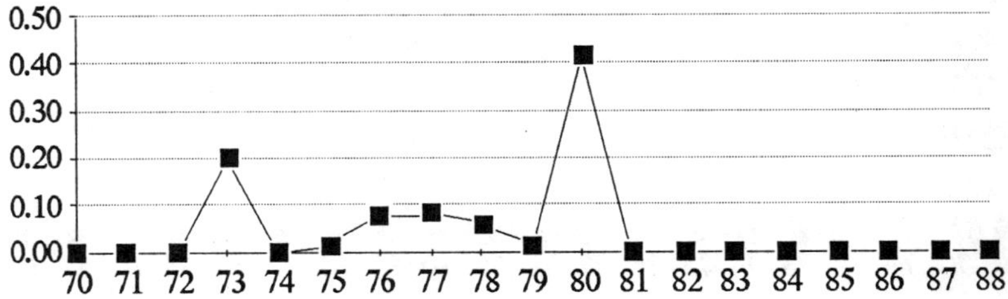

Oceanic Abstracts Records for Portuguese Documents:
Records by Year

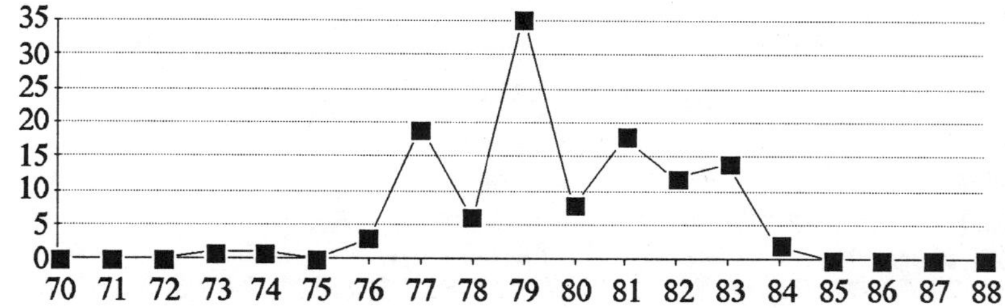

Oceanic Abstracts Records for Portuguese Documents::
Percent of All Records, by Year

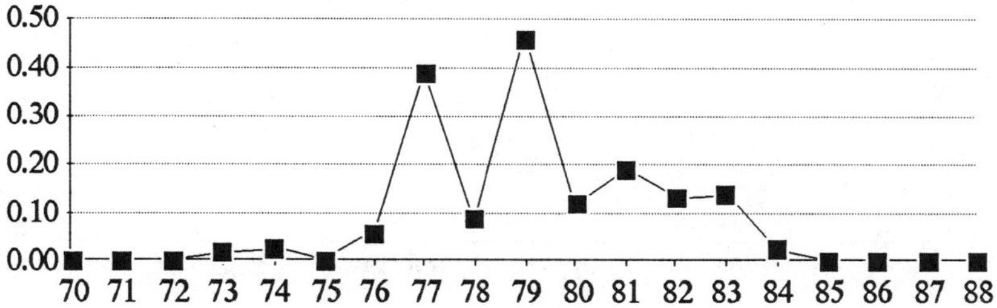

Oceanic Abstracts Records for Russian Documents:
Records by Year

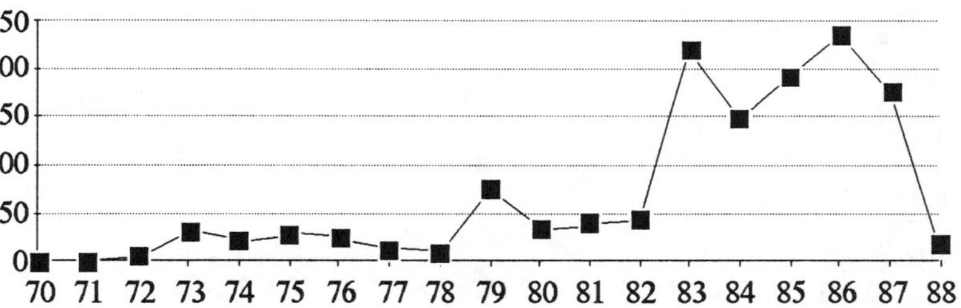

Oceanic Abstracts Records for Russian Documents: Percent of All Records, by Year

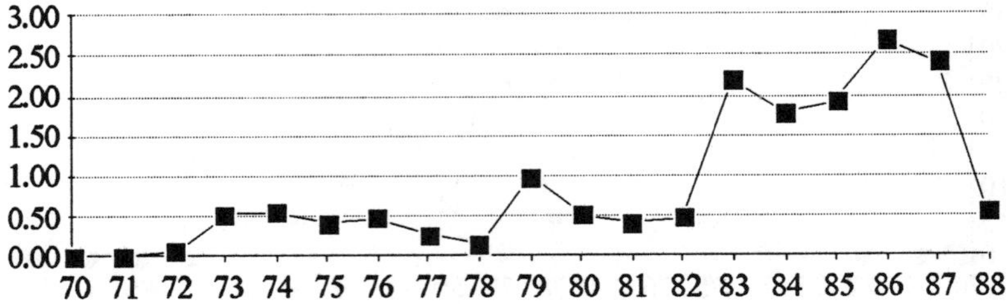

Oceanic Abstracts Records for Spanish Documents: Records by Year

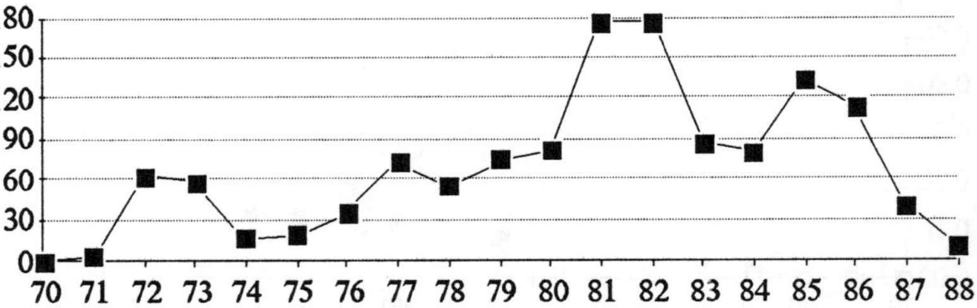

Oceanic Abstracts Records for Spanish Documents: Percent of All Records, by Year

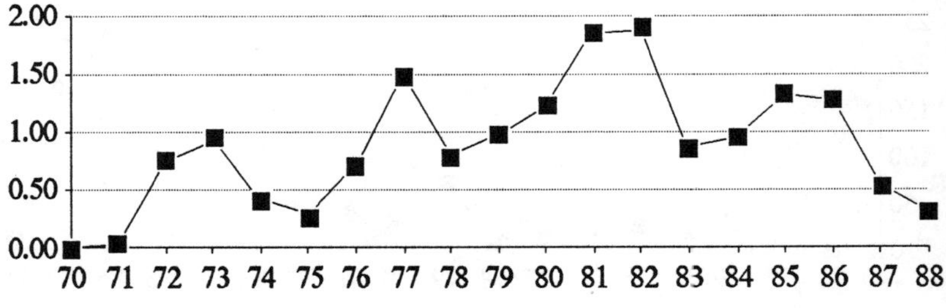

Oceanic Abstracts Summary Languages

Oceanic Abstracts: Records for English Summaries,
Total Records Added, by Year

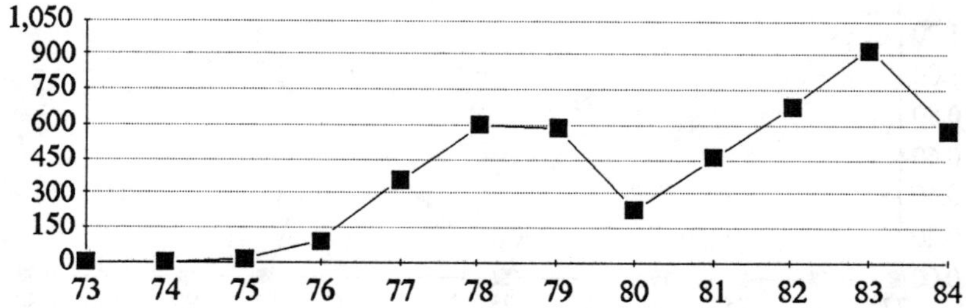

Oceanic Abstracts: Records for English Summaries,
Percent of Summaries Added, by Year

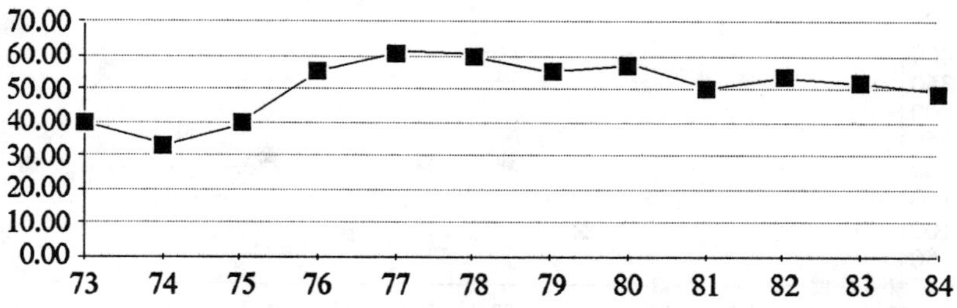

Oceanic Abstracts: Records for Chinese Summaries,
Total Records Added, by Year

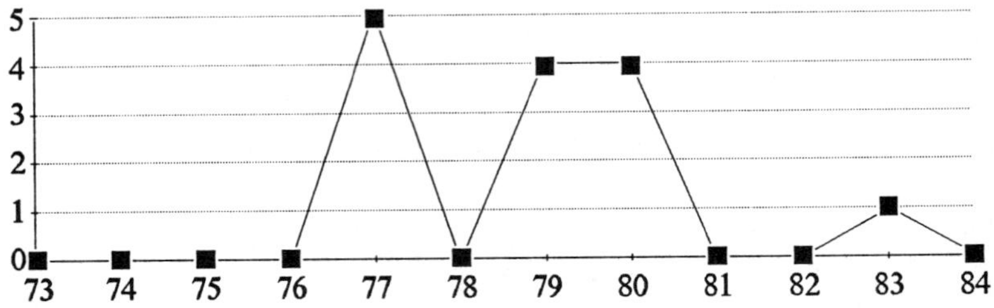

Oceanic Abstracts: Records for Chinese Summaries,
Percent of Total Summaries Added, by Year

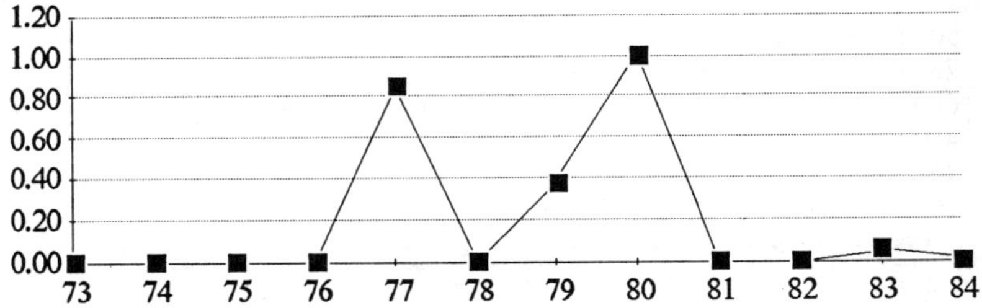

Oceanic Abstracts: Records for French Summaries,
Total Records Added, by Year

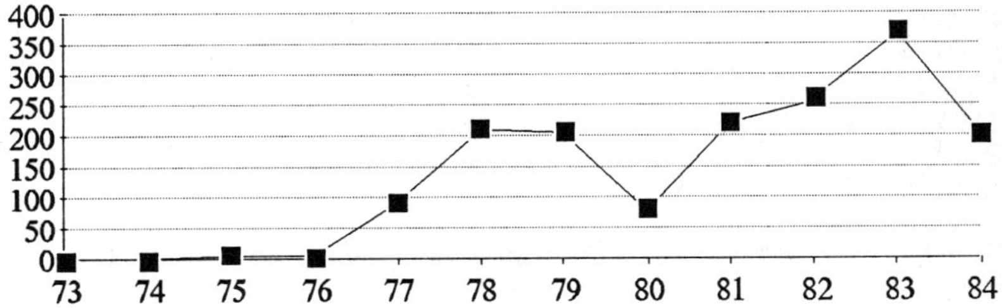

Oceanic Abstracts: Records for French Summaries,
Percent of Summaries Added, by Year

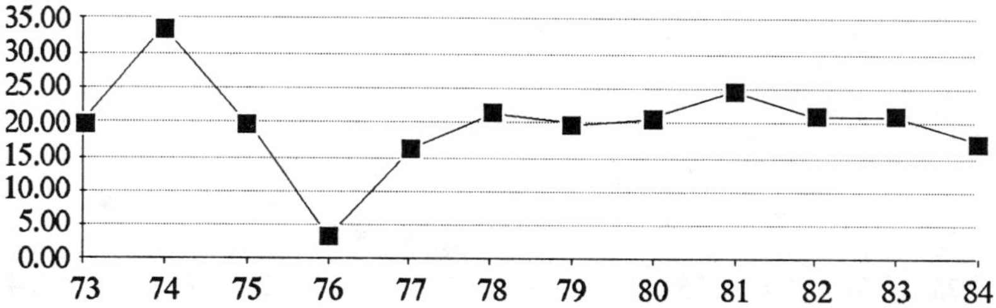

Oceanic Abstracts: Records for German Summaries,
Total Records Added, by Year

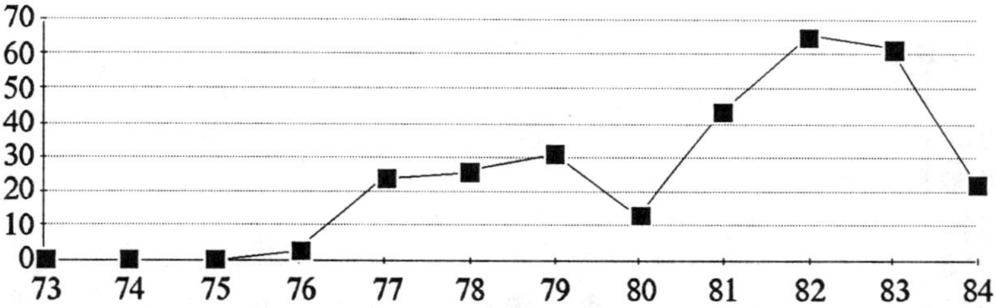

Oceanic Abstracts: Records for German Summaries,
Percent of Total Summaries Added, by Years

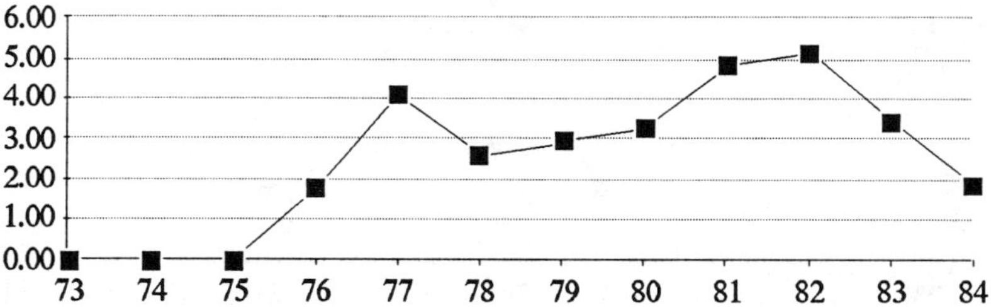

Oceanic Abstracts: Records for Italian Summaries,
Total Records Added, by Year

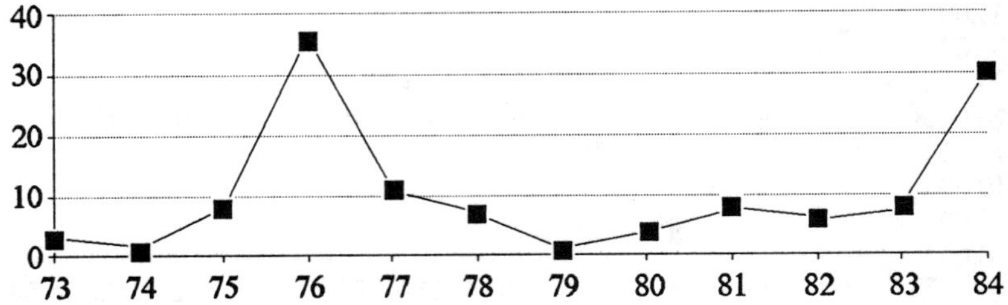

Oceanic Abstracts: Records for Italian Summaries,
Percent of Total Summaries Added, by Year

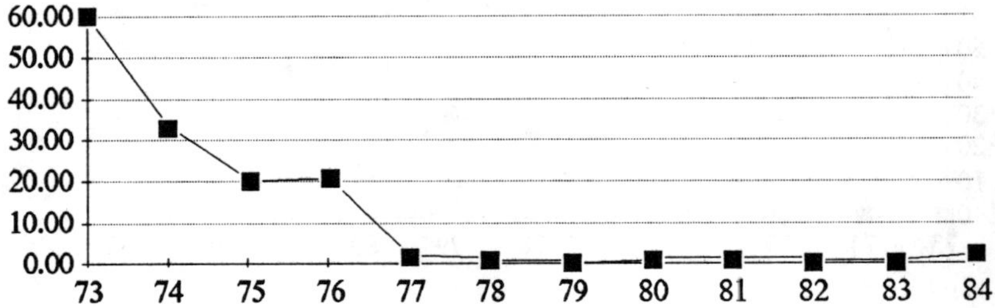

Oceanic Abstracts: Records for Japanese Summaries,
Total Records Added, by Year

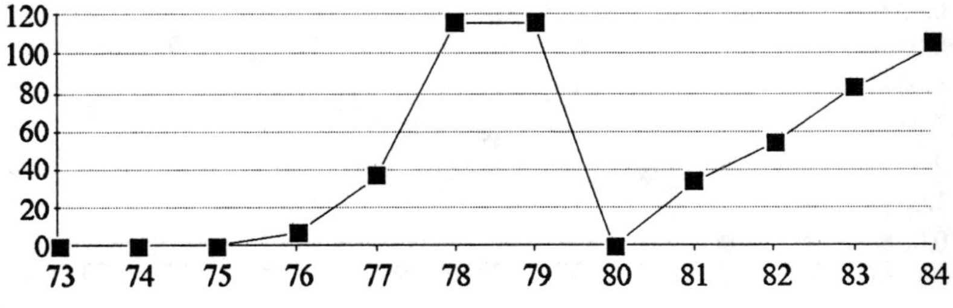

Oceanic Abstracts: Records for Japanese Summaries,
Percent of Summaries Added, by Year

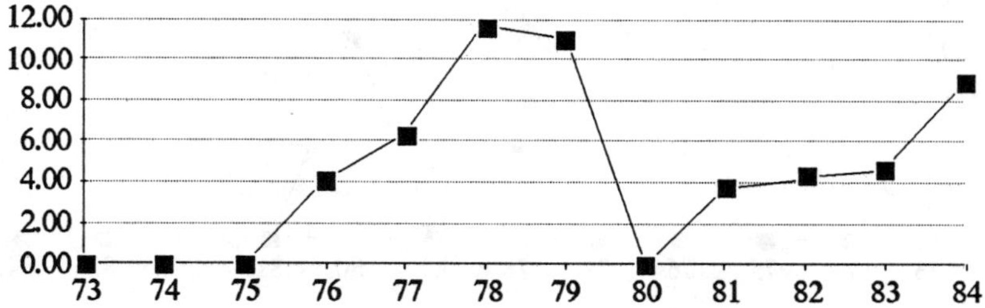

Oceanic Abstracts: Records for Korean Summaries,
Total Records Added, by Year

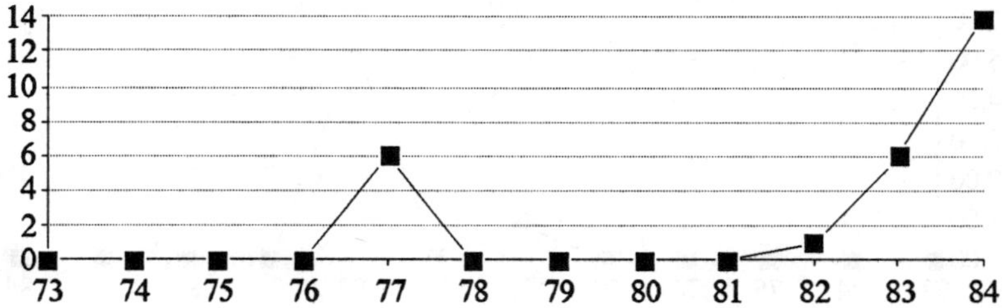

Oceanic Abstracts: Records for Korean Summaries,
Percent of Summaries Added, by Year

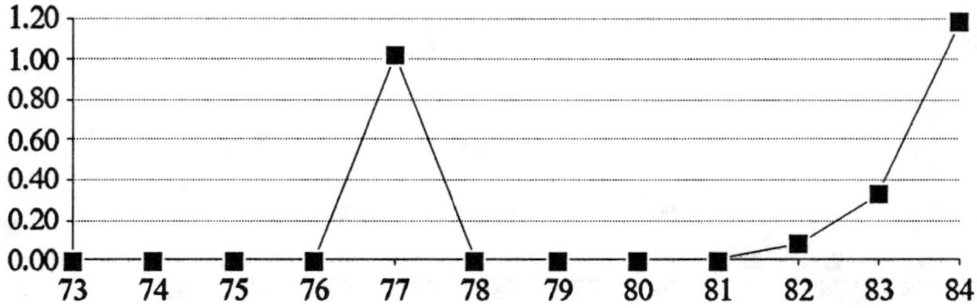

Oceanic Abstracts: Records for Polish Summaries,
Total Records Added, by Year

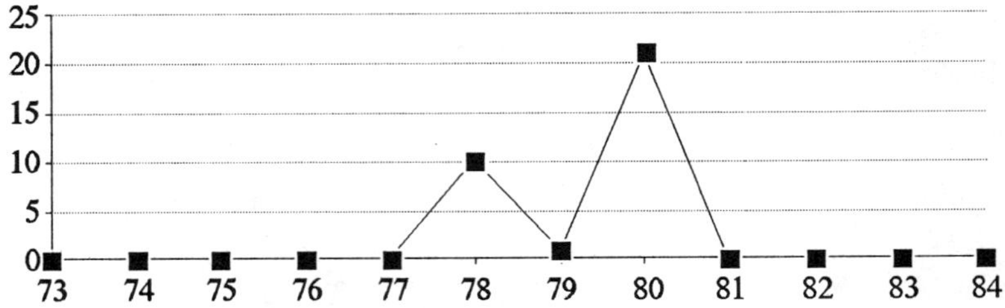

Oceanic Abstracts: Records for Polish Summaries,
Percent of Summaries Added, by Year

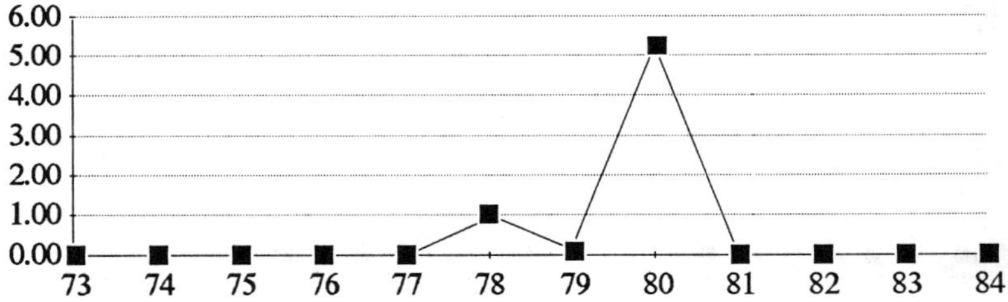

Oceanic Abstracts: Records for Portuguese Summaries,
Total Records Added, by Year

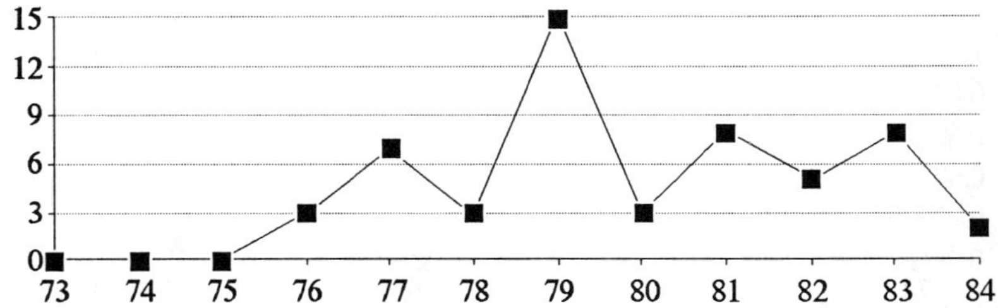

Oceanic Abstracts: Records for Portuguese Summaries,
Percent of Summaries Added, by Year

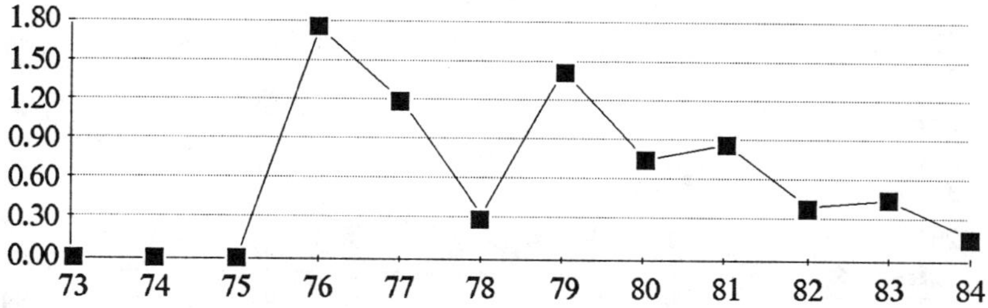

Oceanic Abstracts: Records for Russian Summaries,
Total Records Added, by Year

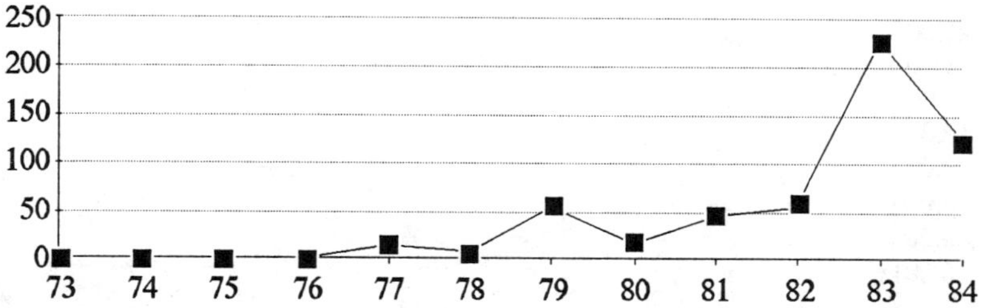

Oceanic Abstracts: Records for Russian Summaries,
Percent of Summaries Added, by Year

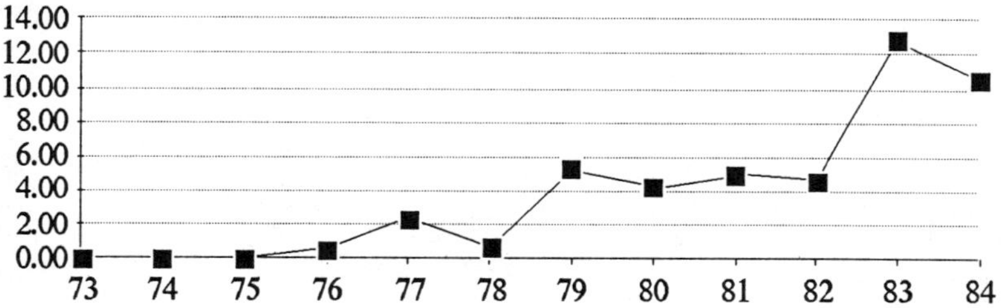

Oceanic Abstracts: Records for Spanish Summaries,
Total Records Added, by Year

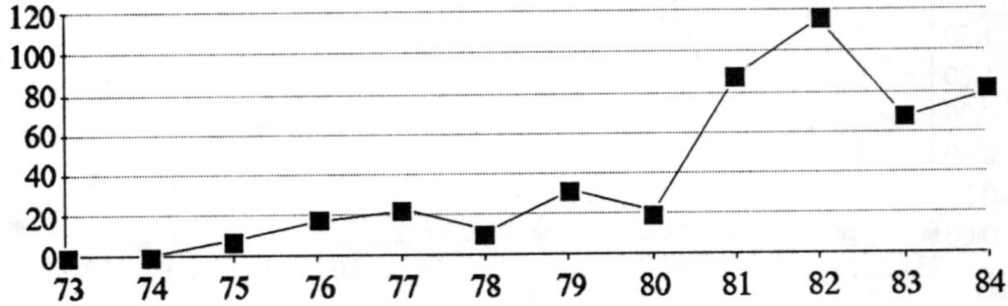

Oceanic Abstracts: Records for Spanish Summaries,
Percent of Summaries Added, by Year

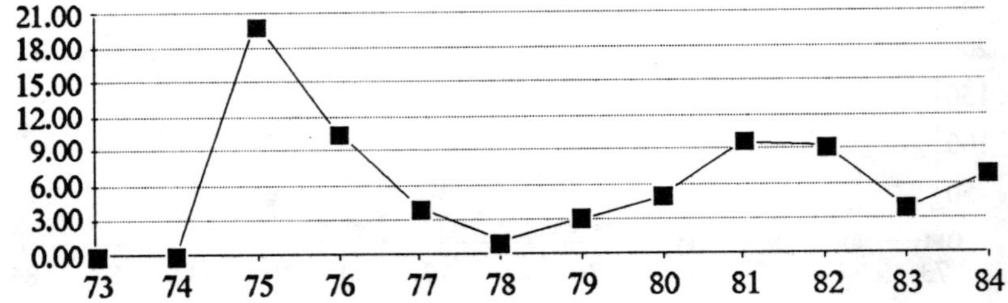

Public Affairs Information Service (PAIS)

The following charts show the trends found for languages in the PAIS file. The English charts are presented first, followed by the other languages in alphabetical order.

Percentages are whole percents: that is, ".65%" indicates one sixty-ffive hundredth of a percent, "1.5%" indicates one and a half percent.

PAIS Records for English Documents:
Records by Year

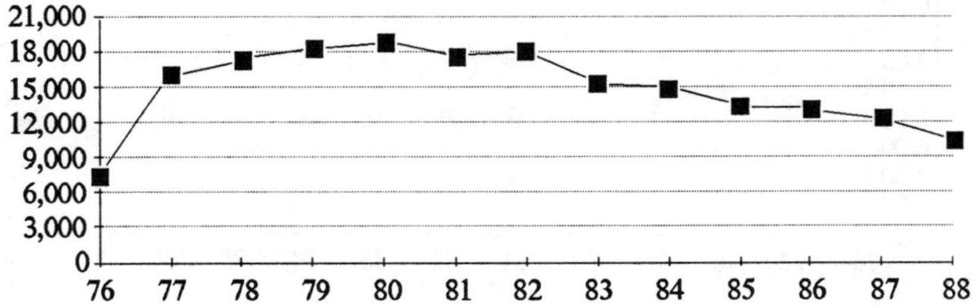

PAIS Records for English Documents:
Percent of All Records, by Year

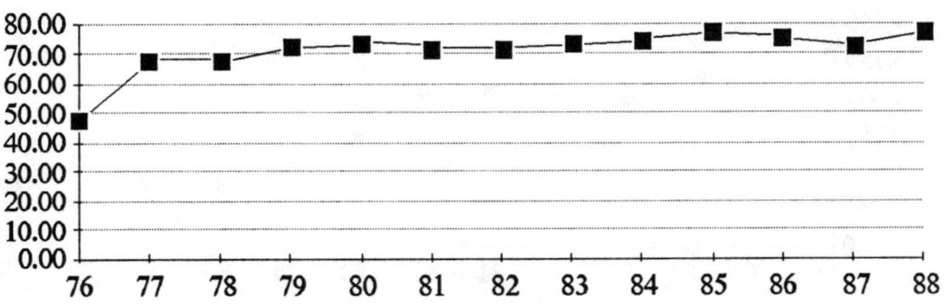

PAIS Records for French Documents:
Records by Year

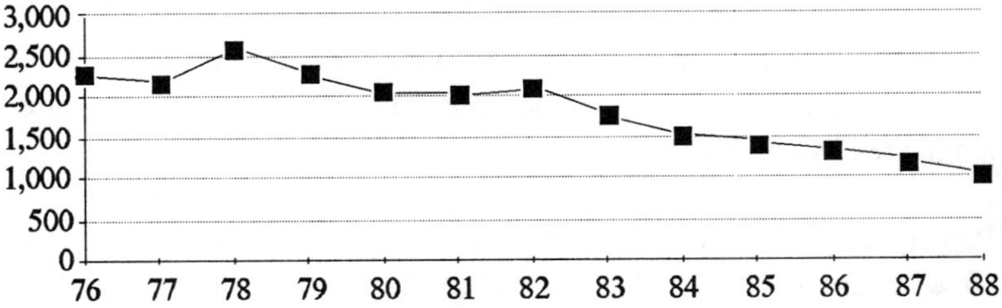

PAIS Records for French Documents:
Percent of All Records, by Year

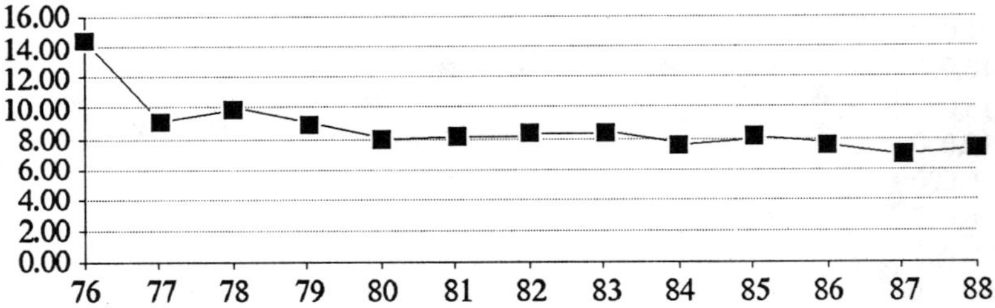

PAIS Records for German Documents:
Records by Year

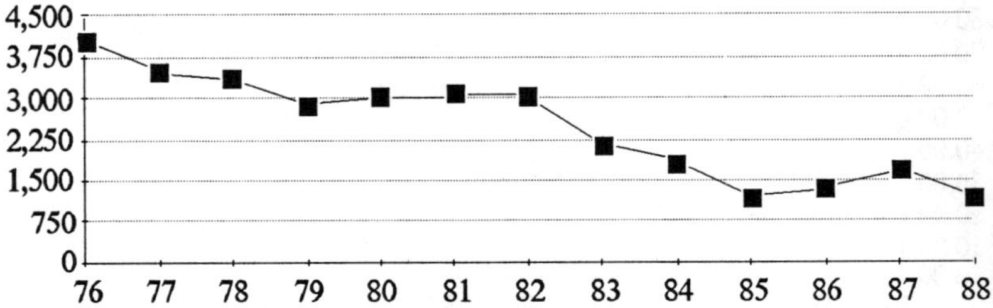

PAIS Records for German Documents:
Percent of All Records, by Year

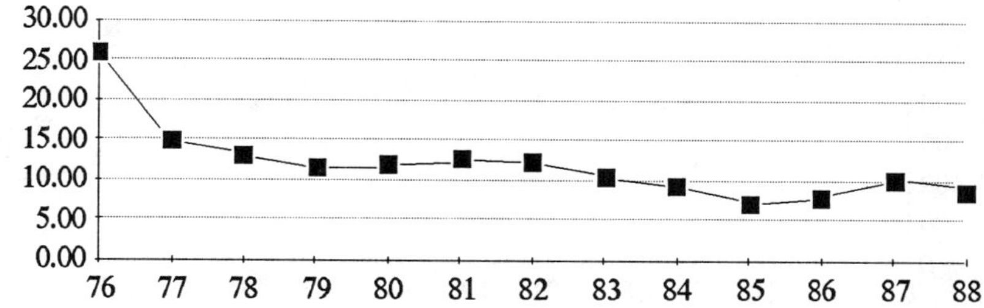

PAIS Records for Italian Documents:
Records by Year

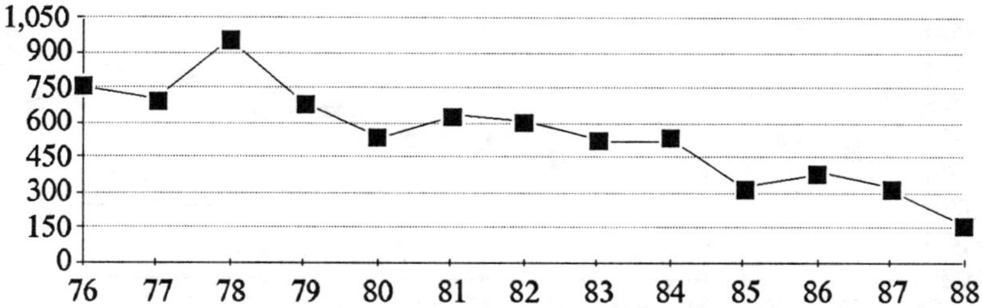

PAIS Records for Italian Documents:
Percent of All Records, by Year

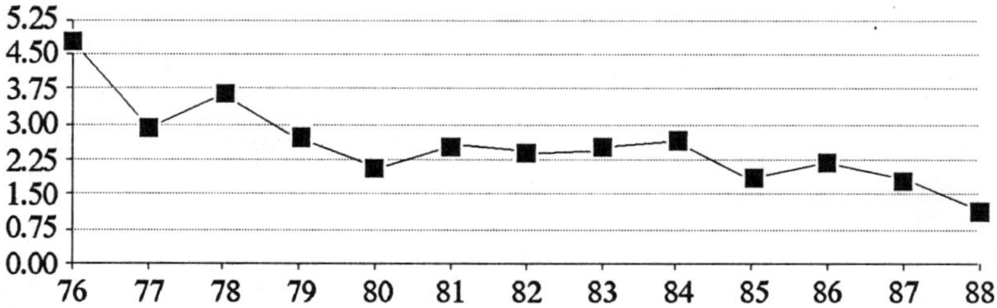

PAIS Records for Spanish Documents:
Records by Year

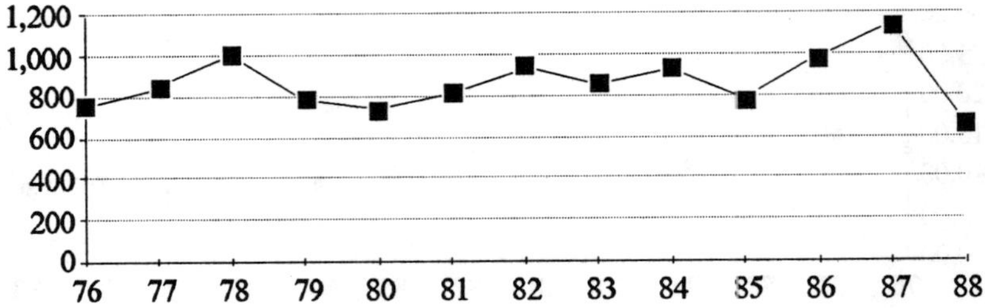

PAIS Records for Spanish Documents:
Percent of All Records, by Year

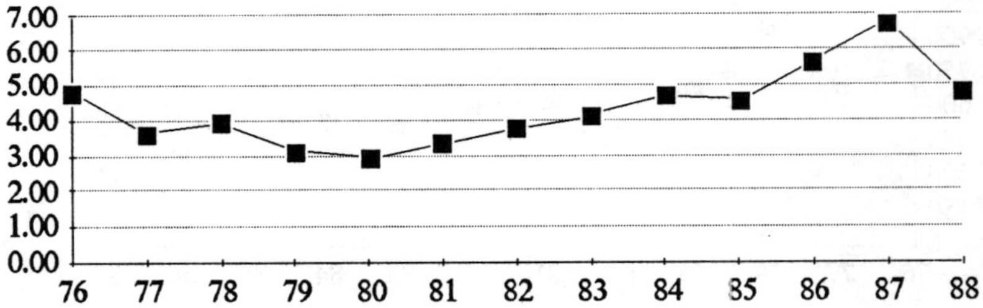

PsycInfo Records

The following charts show the trends found for languages in the PsycInfo file. The English charts are presented first, followed by the other languages in alphabetical order.

Percentages are whole percents: that is, ".65%" indicates one sixty-five hundredth of a percent, "1.5%" indicates one and a half percent.

PsycInfo Records for English Documents:
Records by Year

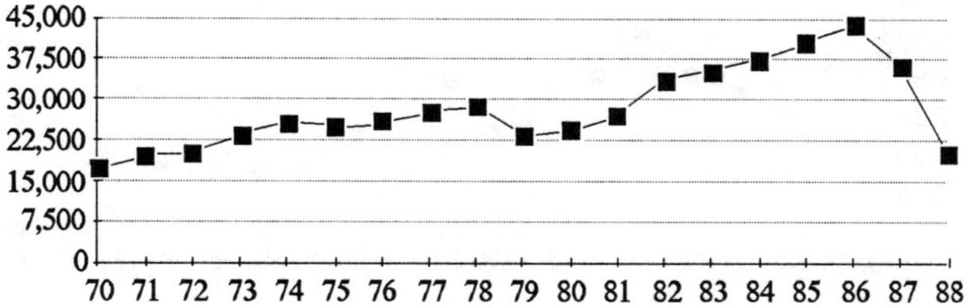

PsycInfo Records for English Documents:
Percent of All Records, by Year

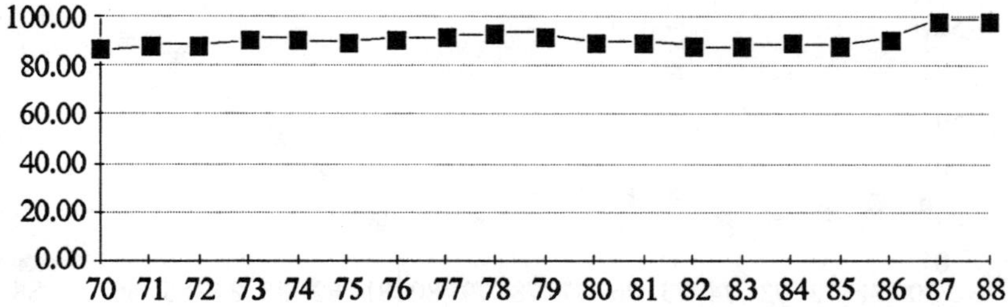

PsycInfo Records for Czech Documents:
Records by Year

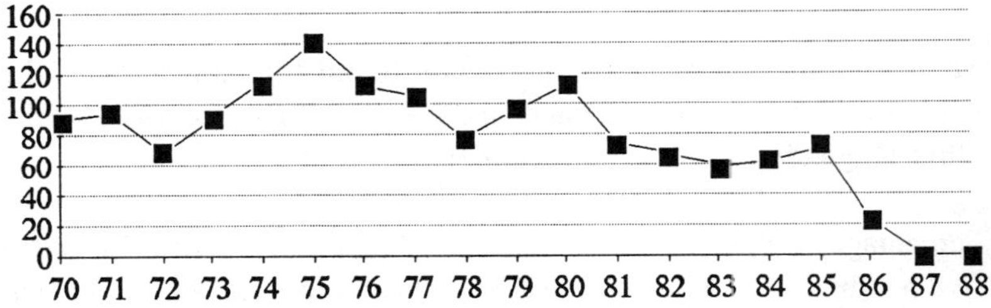

PsycInfo Records for Czech Documents:
Percent of All Records, by Year

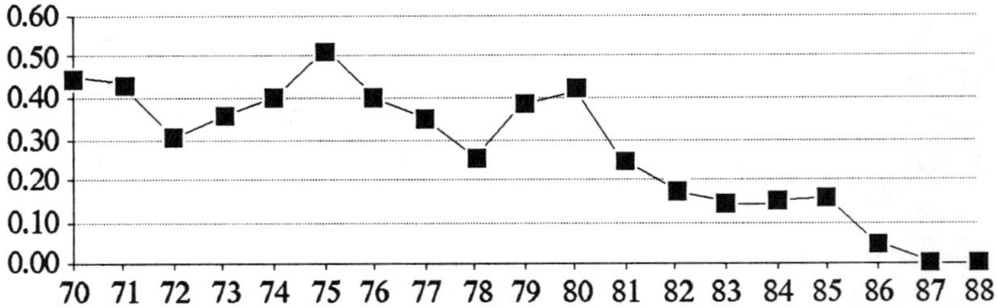

PsycInfo Records for Dutch Documents:
Records by Year

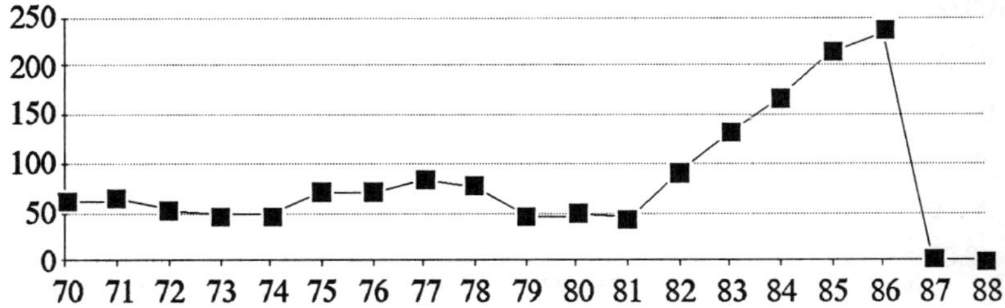

PsycInfo Records for Dutch Documents:
Percent of All Records, by Year

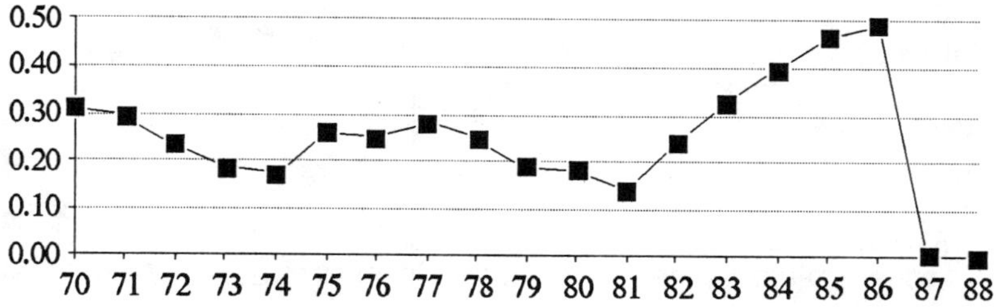

PsycInfo Records for French Documents:
Records by Year

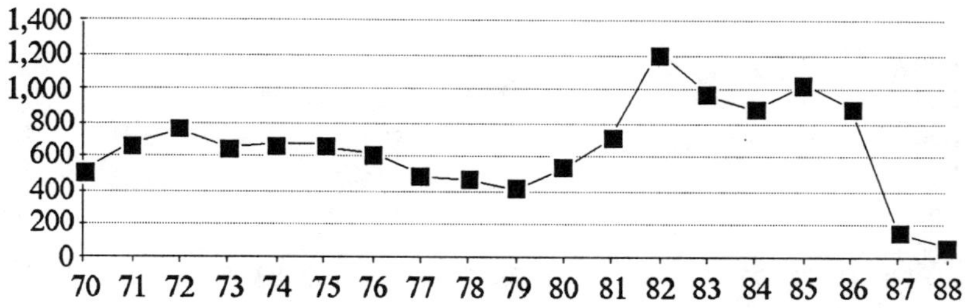

PsycInfo Records for French Documents:
Percent of All Records, by Year

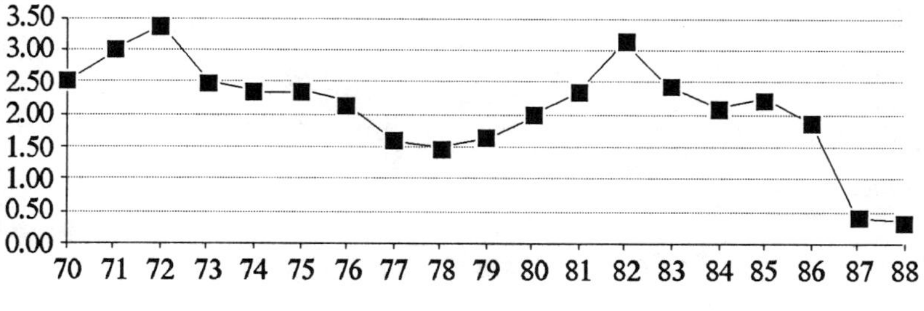

PsycInfo Records for German Documents:
Records by Year

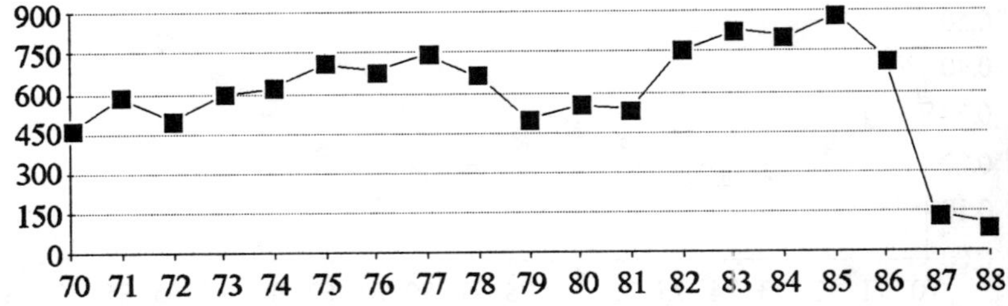

PsycInfo Records for German Documents:
Percent of All Records, by Year

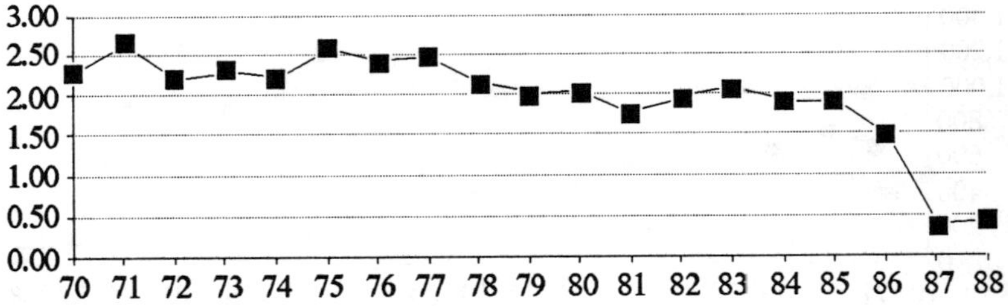

PsycInfo Records for Italian Documents:
Records by Year

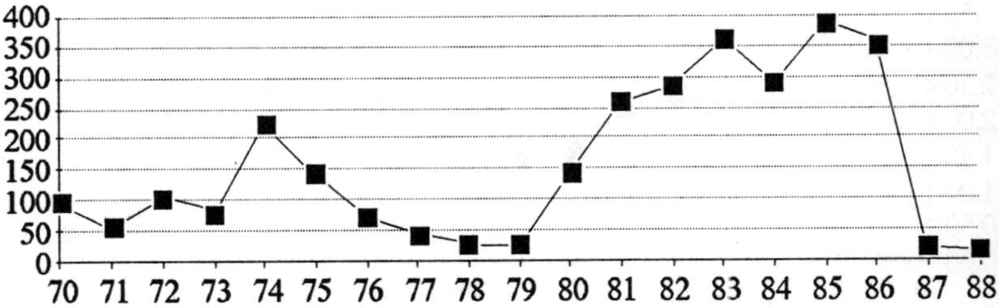

PsycInfo Records for Italian Documents:
Percent of All Records, by Year

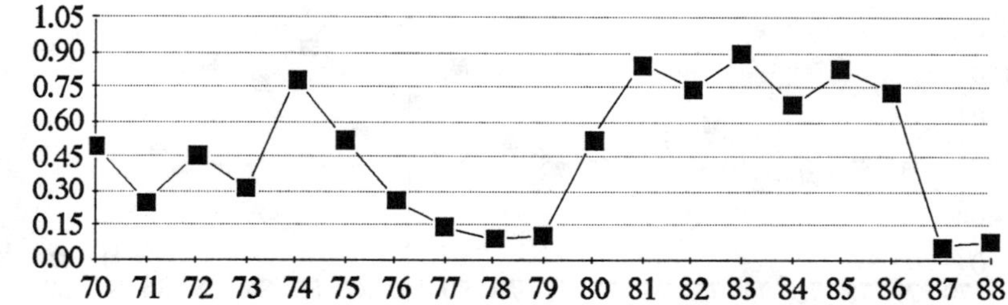

PsycInfo Records for Japanese Documents:
Records by Year

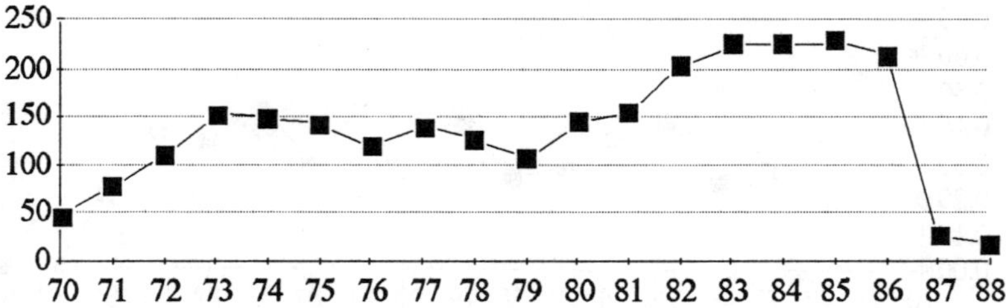

PsycInfo Records for Japanese Documents:
Percent of All Records, by Year

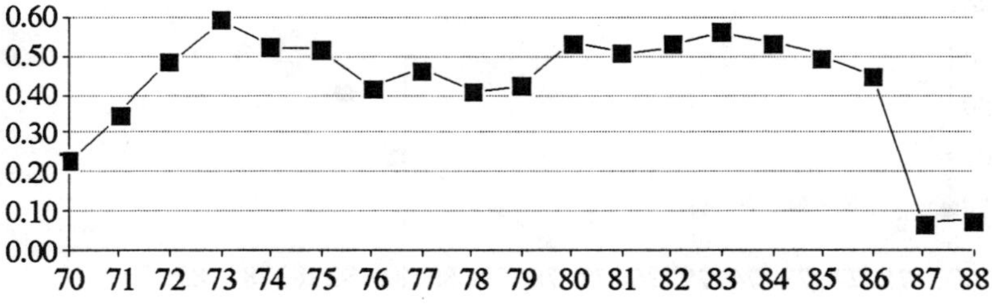

PsycInfo Records for Polish Documents:
Records by Year

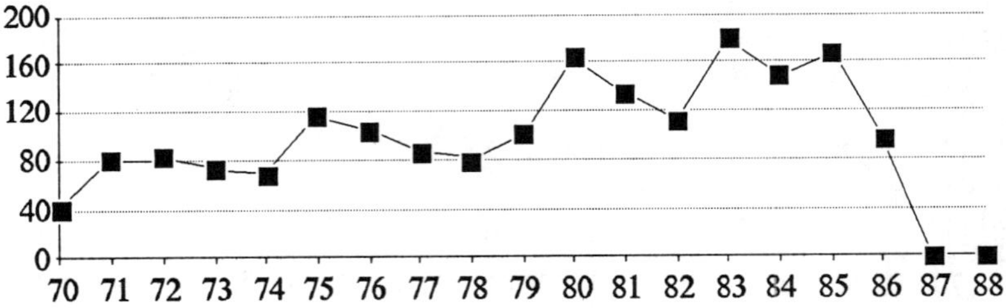

PsycInfo Records for Polish Documents:
Percent of All Records, by Year

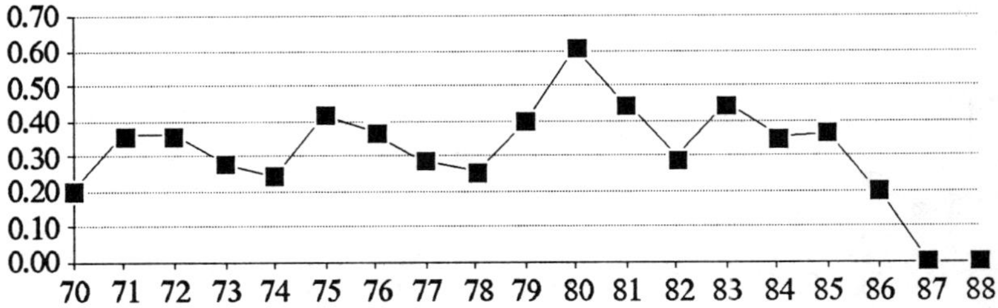

PsycInfo Records for Portuguese Documents:
Records by Year

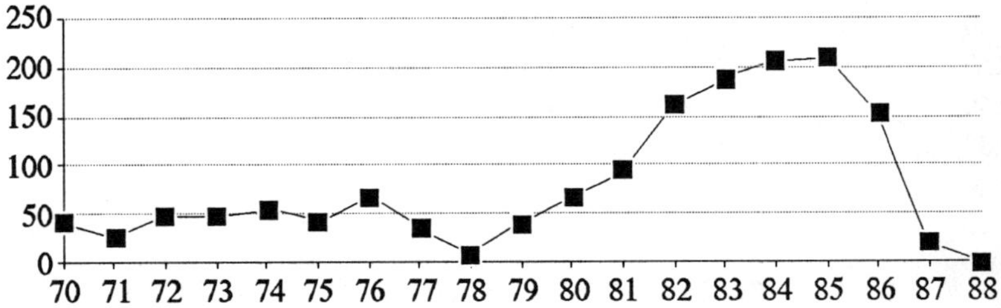

PsycInfo Records for Portuguese Documents:
Percent of All Records, by Year

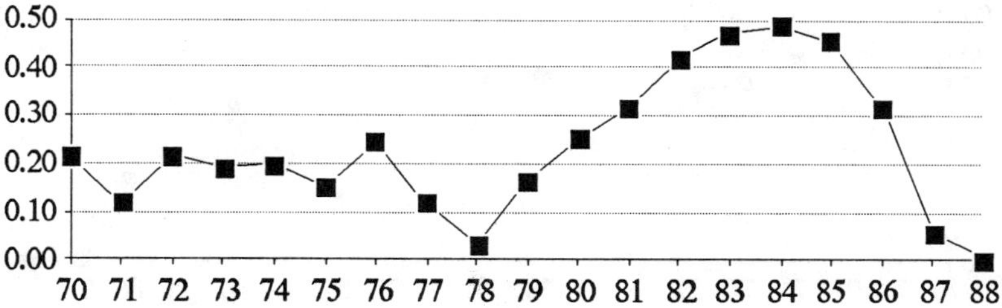

PsycInfo Records for Russian Documents:
Records by Year

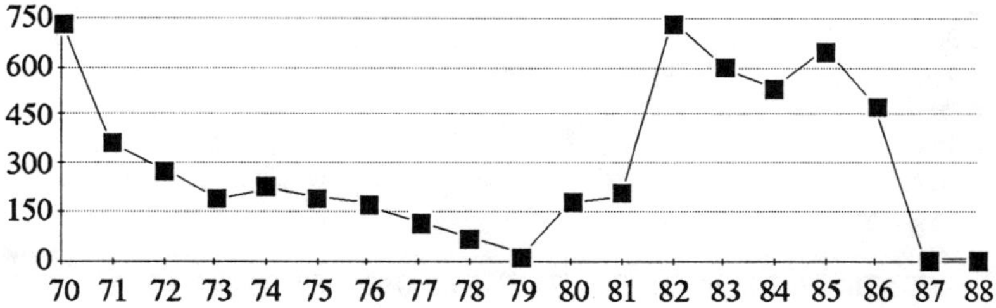

PsycInfo Records for Russian Documents:
Percent of All Records, by Year

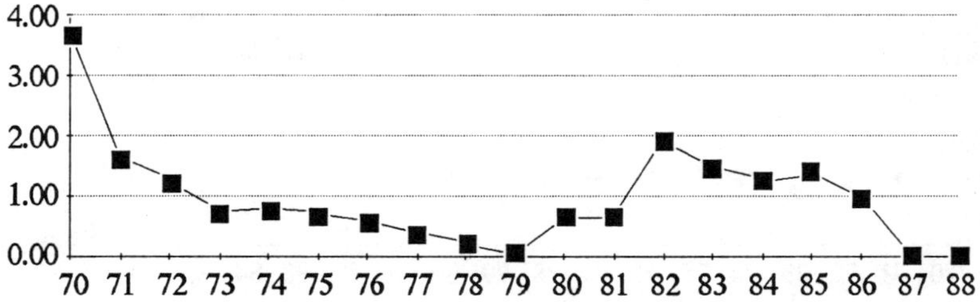

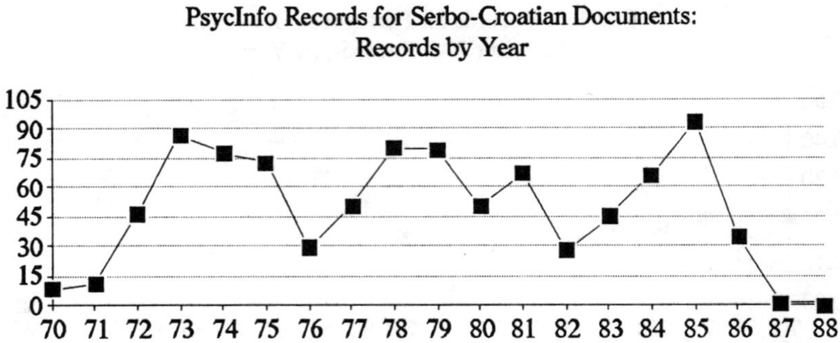
PsycInfo Records for Serbo-Croatian Documents:
Records by Year
105
90
75
60
45
30
15
0
70 71 72 73 74 75 76 77 78 79 80 81 82 83 84 85 86 87 88

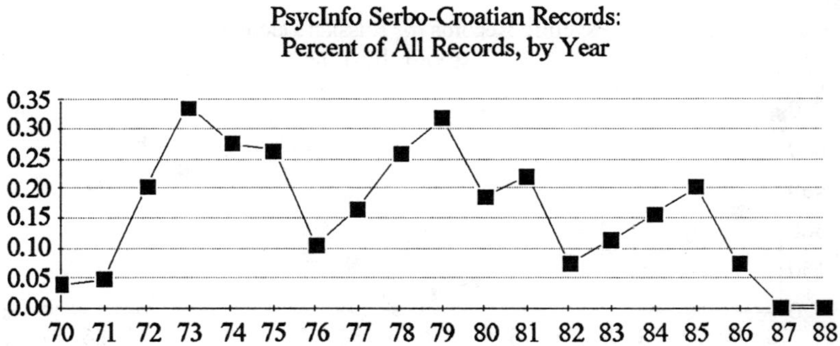
PsycInfo Serbo-Croatian Records:
Percent of All Records, by Year
0.35
0.30
0.25
0.20
0.15
0.10
0.05
0.00
70 71 72 73 74 75 76 77 78 79 80 81 82 83 84 85 86 87 88

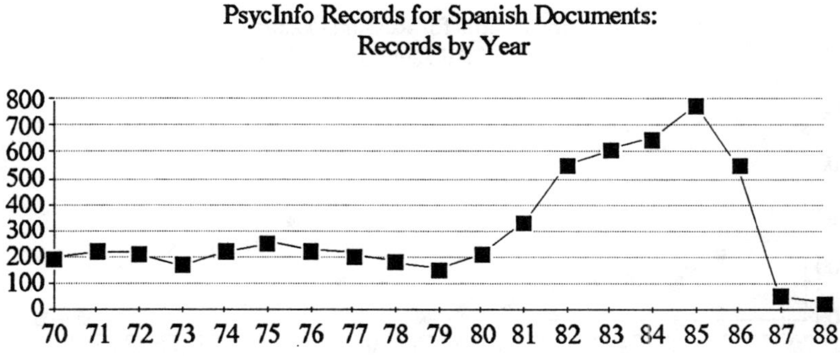
PsycInfo Records for Spanish Documents:
Records by Year
800
700
600
500
400
300
200
100
0
70 71 72 73 74 75 76 77 78 79 80 81 82 83 84 85 86 87 88

PsycInfo Records for Spanish Documents:
Percent of All Records, by Year

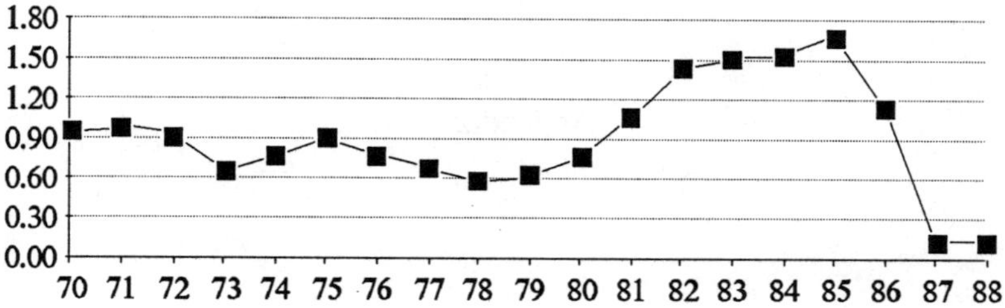

Appendix 4

The Unesco Statistics

The charts for the Unesco data are presented in the following order: (1) all statistics reported in the given language, (2) statistics showing number of documents reported by the selected, frequently-reporting countries, (3) low estimates of production, (4) medium estimates, (5) high estimates, and (6) the number of countries actually reporting data for the period of study.

UNESCO Statistics: All Countries,
Documents Reported in English, by Year

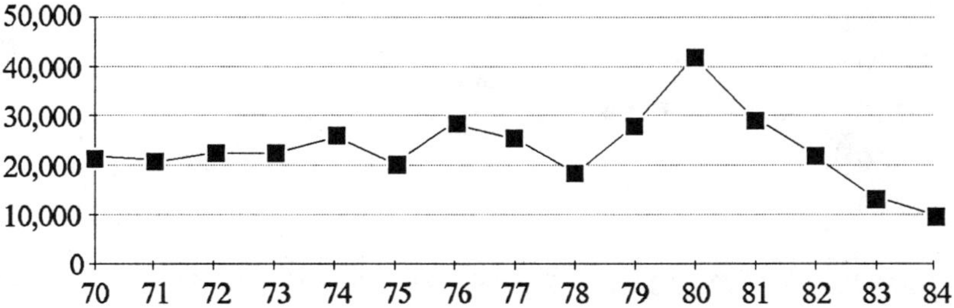

Unesco English Document Statistics:
Missing Data Filled in by Low Estimates

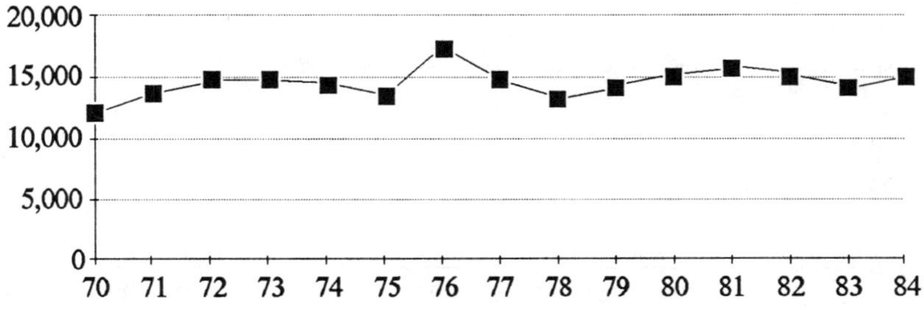

Unesco English Document Statistics,
Missing Data Filled in by Medium Estimates

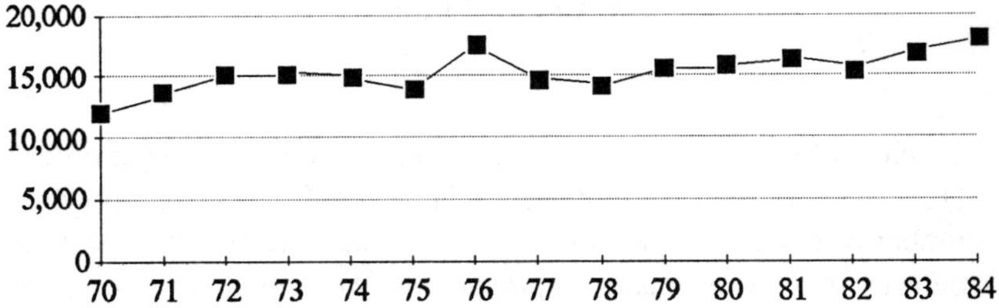

Unesco English Document Statistics:
Missing Data Filled in by High Estimates

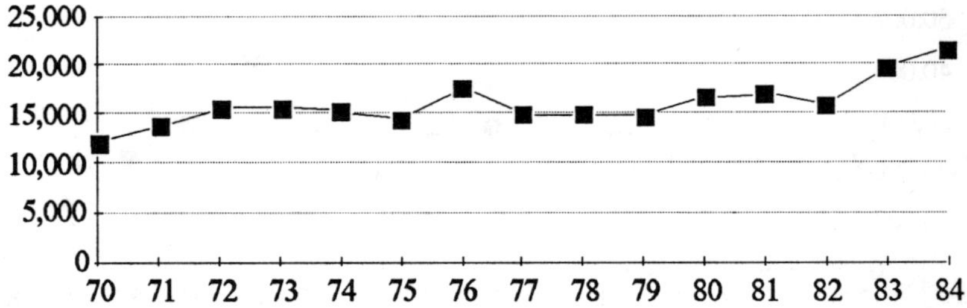

Unesco English Document Statistics,
Number of Countries Reporting, by Year

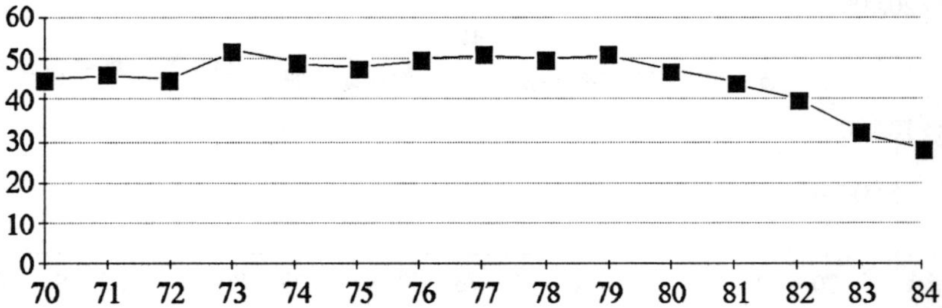

Unesco Statistics: All Countries,
Documents Reported in French, by Year

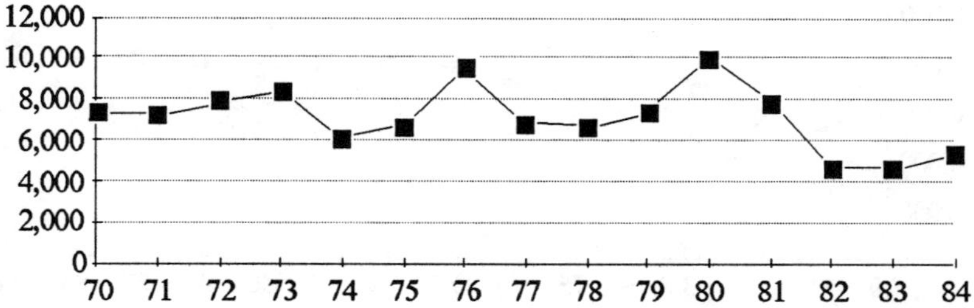

Unesco French Document Statistics:
Missing Data Filled in by Low Estimates

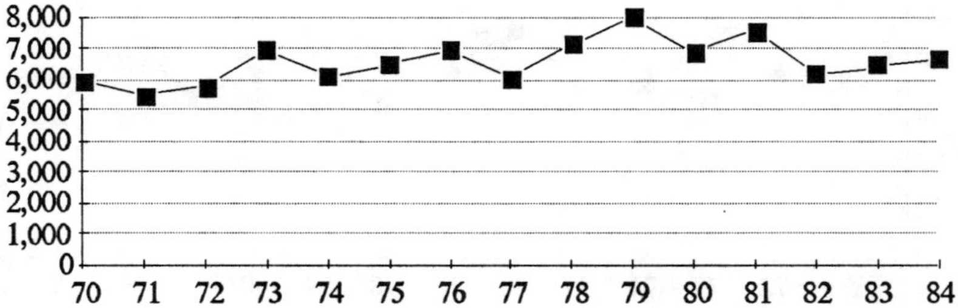

Unesco French Document Statistics:
Missing Data Filled in by Medium Estimates

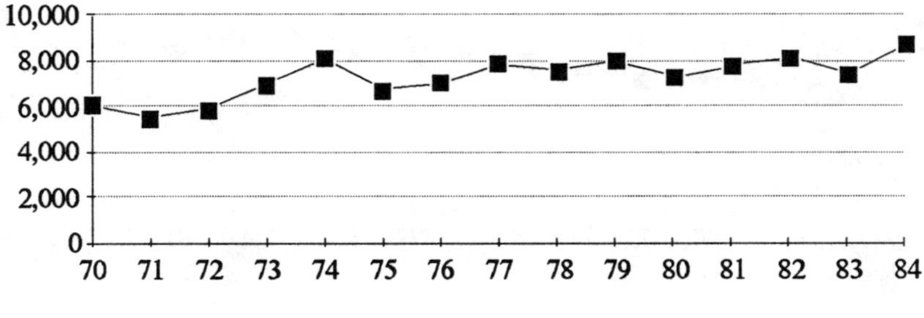

Unesco French Document Statistics:
Missing Data Filled in by High Estimates

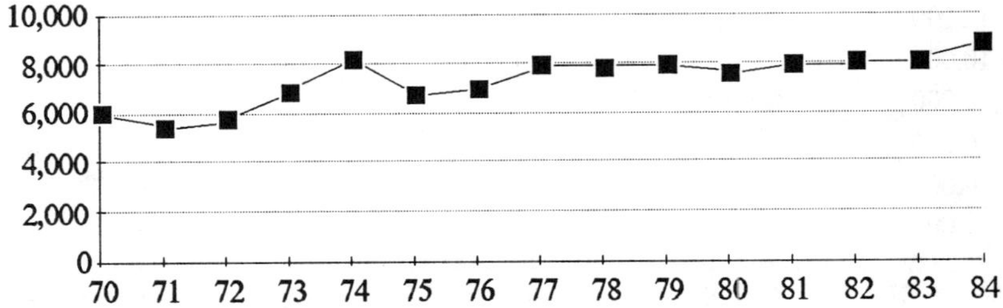

Unesco French Document Statistics,
Number of Countries Reporting, by Year

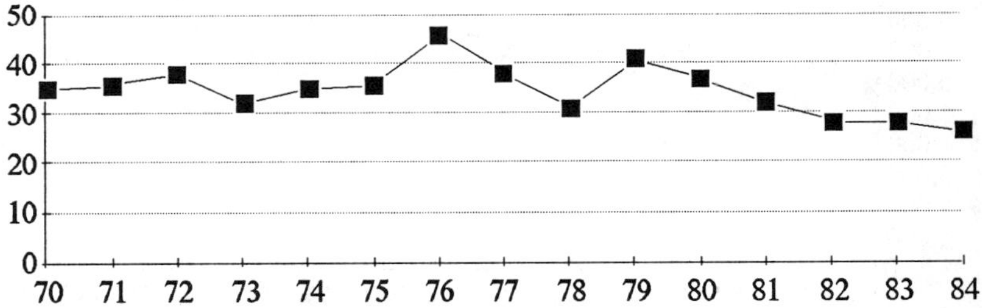

Unesco Statistics: All Countries,
Documents Reported in German, by Year

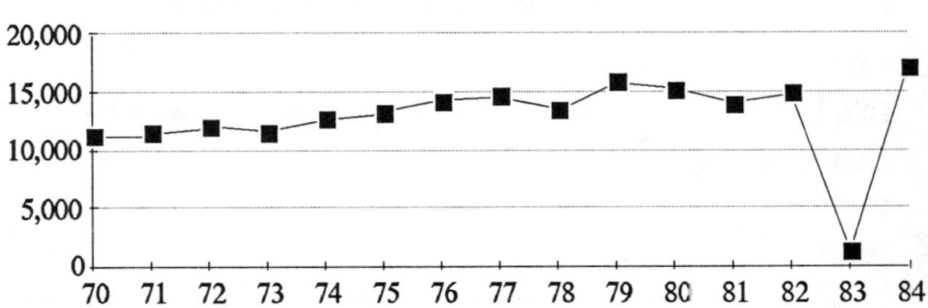

Unesco German Document Statistics:
Missing Data Filled in by Low Estimates

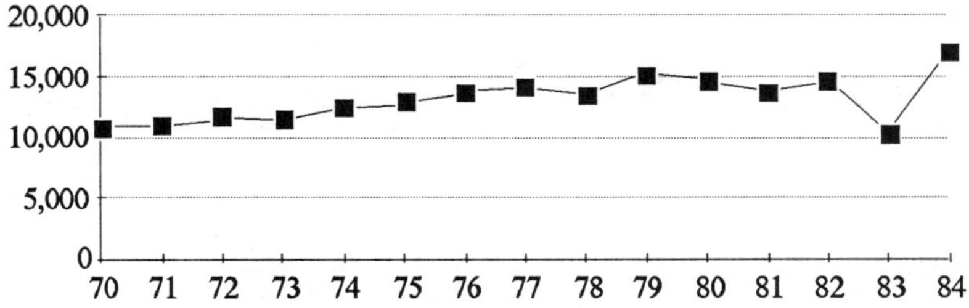

Unesco German Document Statistics:
Missing Data Filled in by Medium Estimates

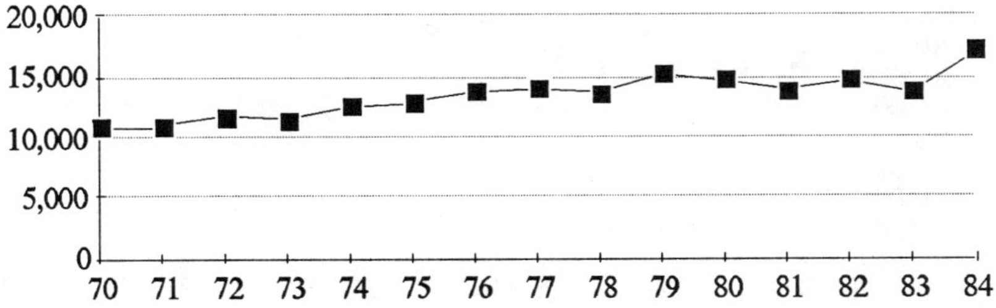

Unesco German Document Statistics:
Missing Data Filled in by High Estimates

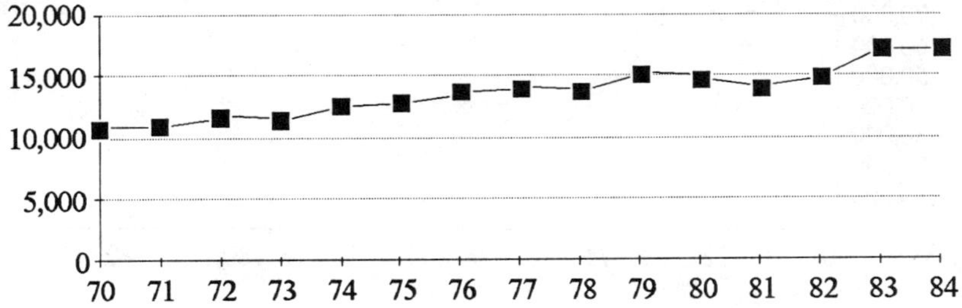

Unesco German Document Statistics,
Number of Countries Reporting, by Year

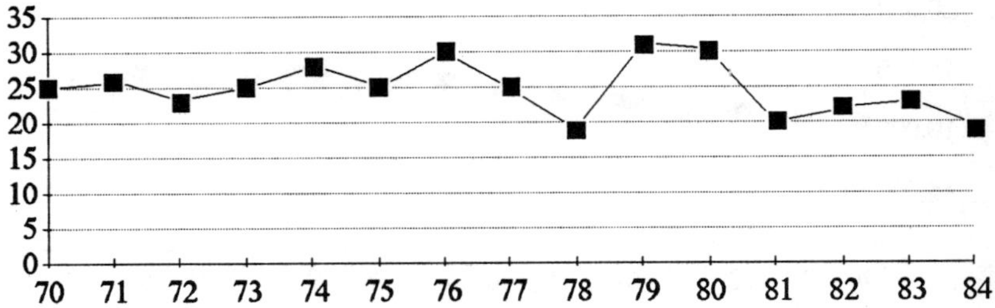

Unesco Statistics: All Countries Reporting,
Documents Reported in Russian, by Year

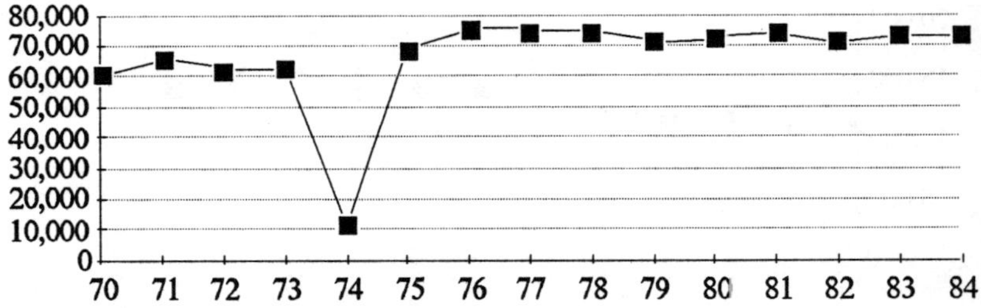

Unesco Russian Document Statistics:
Missing Data Filled in by Low Estimates

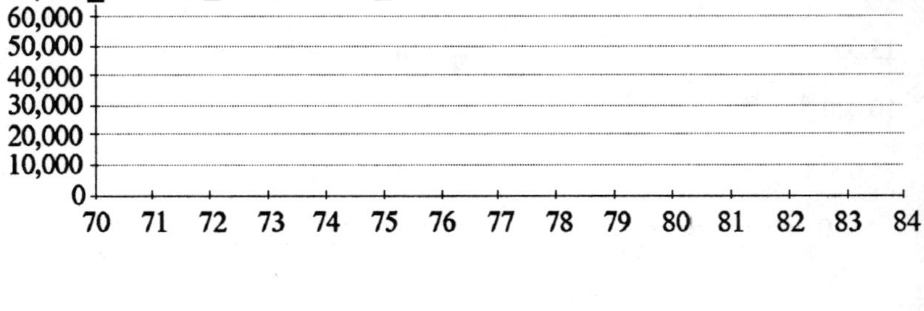

Unesco Russian Document Statistics:
Missing Data Filled in by Medium Estimates

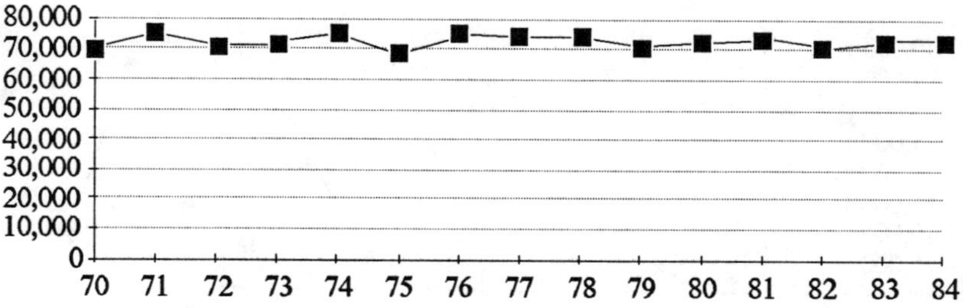

Unesco Russian Document Statistics:
Missing Data Filled in by High Estimates

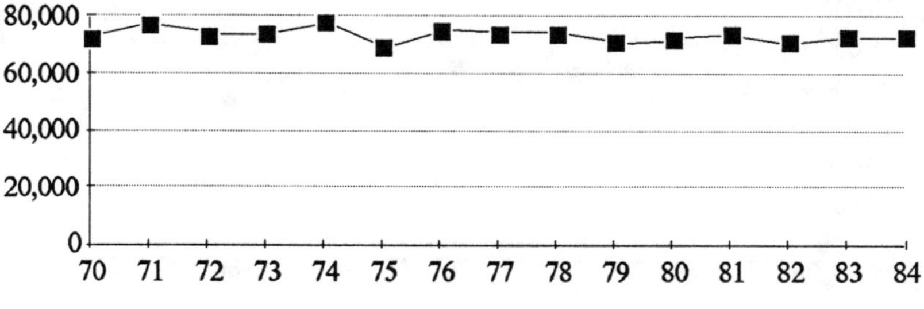

Unesco Russian Document Statistics,
Number of Countries Reporting, by Year

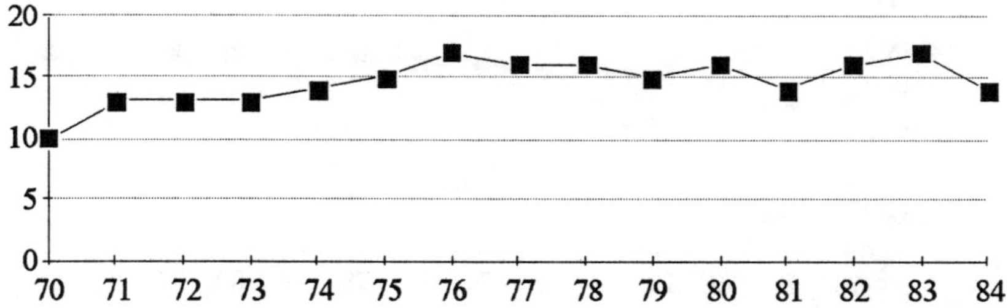

Unesco Statistics: All Countries Reporting,
Documents Reported in Spanish, by Year

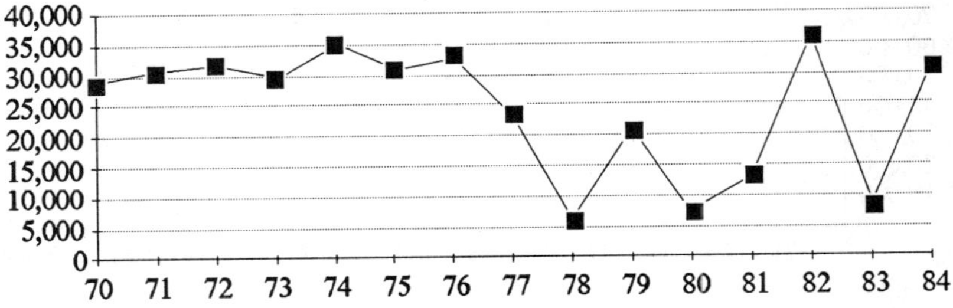

Unesco Spanish Document Statistics:
Missing Data Filled in by Low Estimates

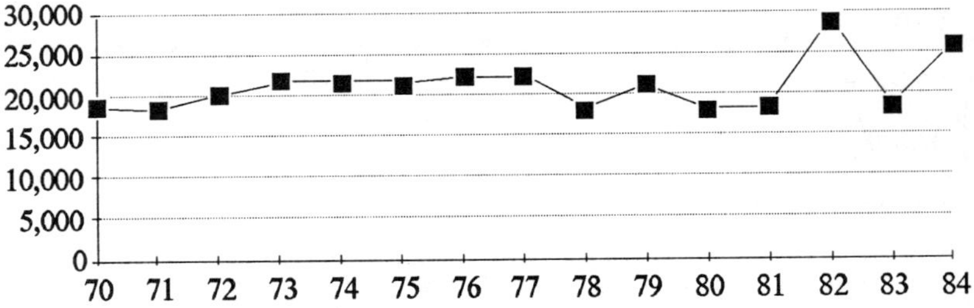

Unesco Spanish Document Statistics:
Missing Data Filled in by Medium Estimates

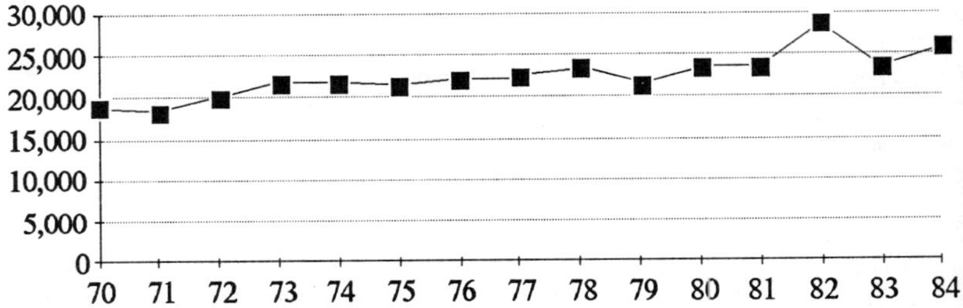

Unesco Spanish Document Statistics:
Missing Data Filled in by High Estimates

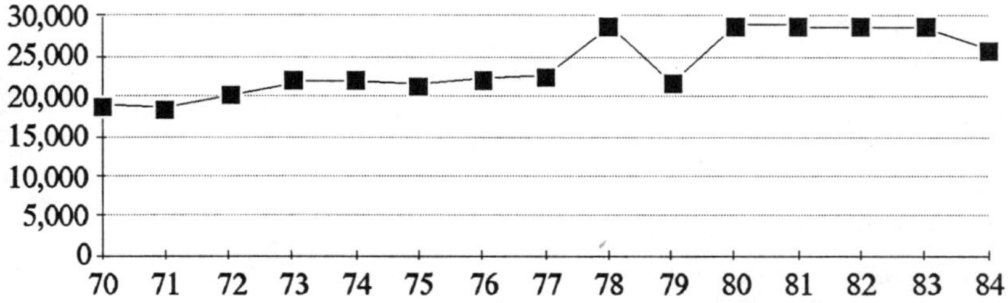

Unesco Spanish Document Statistics,
Number of Countries Reporting, by Year

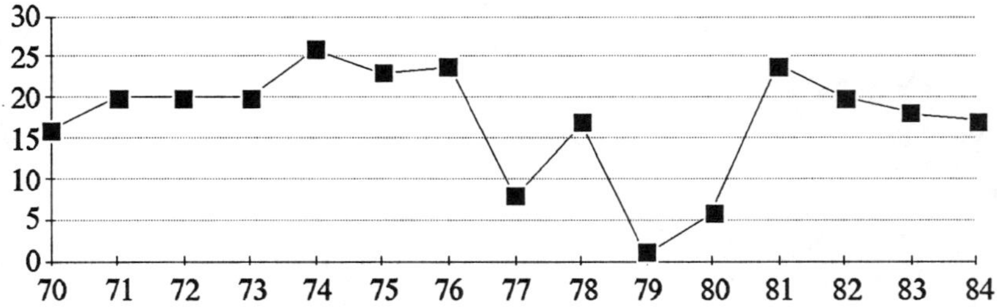

Appendix 5

NOTES ON METHODOLOGY

The data were collected for all but one database by searching the various fields present in the DIALOG files. The DT (Document Type, LA (Language), CP (Country of Publication) and PY (Year of Publication) fields were used. If a field (such as the country of publication field) was not present, no further effort was made to search or process the data.

The languages analyzed were those used in the databases, without extensive standardization. Obvious typographical errors (such as "Gemran" for "German") were corrected and counted appropriately. The data are presented as they appear in the databases, and only minimal standardization has been done across the files. For example, Iranian, Persian, and Farsi are all essentially the same language. However, if a database used all three values as if they were separate, then all three were analyzed as if they were separate. While this inconsistency does skew the results for some languages, these languages are not the major languages in any of the files. The same can be said for country of publication, and for document type. For example, one file uses the term "journal article" and the other uses "serial."

Some databases included obvious extraneous information in some fields. This was just ignored.

The Oceanic Abstracts file was analyzed on an experimental basis by a program written in the MIDAS programming language on the University of Michigan mainframe. For cost and time reasons, a simpler approach was used for the other databases.

In the 1986 Oceanic Abstracts file, 554 records contained Summary Language (NT) fields, and 54 of these fields contained only extraneous information such as illustration notes, series statements, ISBN's, and notes on materials included such as chapters and articles. At least one contained the summary language as well as other extraneous data such as the NTIS order number. These inconsistencies were handled by constructing the analysis program to find only the genuine languages.

The analysis of databases in Chapter 3 was accomplished by construction of a Pro-Cite database containing entries for each database, and performing various searches and sorts. The analysis of the DIALOG file was accomplished by on-line searches of the appropriate fields.

BIBLIOGRAPHY

BIBLIOGRAPHY

Adams, Scott. *Foreign users of U. S. bibliographic databases in biology and medicine.* **Library Trends**; July 1974; 23(1): 153-163.

Aborn, Peter K. [and others]. Supplying machine-readable databases to users outside the United States. In: **Proceedings of the American Society for Information Science**; 1973. Washington, D.C.: ASIS; 1973: 97-103.

Aiyepeku, Wilson O. *The languages and format of geographical literature: a comparative study.* **International Library Review**; Jan. 1973; 5(1): 53-62.

Anderson, J.D. Foreign language barriers in information transfer. **Journal of Education for Librarianship**; 1974; 14(3): 171-84.

Antoni, Professor. University of Milan, Italy. Private communication, 1985.

Bailey, Richard W; Görlach, Manfred. **English as a World Language**. Ann Arbor, MI: Univ. of Michigan Press; 1982.

Baker, Dale B. *Recent trends in growth of chemical literature.* **Chemical and Engineering News**; May 1976; 54(20): 23-27.

Baldauf, R. B.; Jernudd, B. H. *Language use patterns in the fisheries periodical literature.* **Scientometrics**; July 1983; 5(4): 245-255.

Bazerman, Charles. *Scientific writing as a social act: A review of the literature of the sociology of science.* Anderson, Paul V.; Brockman, R. John; Miller, Carolyn R.; Eds. **New Essays in Technical and Scientific Communication: Research, Theory, Practice.** Farmingdale, NY: Baywood, 1983.

Bernal, J. D. **The Social Function of Science.** Cambridge, MA: The M. I. T. Press; 1939.

Boas, R. P. Award for Distinguished Service to Otto Neugebauer. **American Mathematical Monthly**; Feb. 1979; 86(2): n.p.

Boffey, Philip M. *Officials seek to stem flow of scientific data to Russia.* **New York Times**; Jan. 10, 1982: Y13, Col. 1.

Bolle, Sonja. *Foreign language makes a comeback.* **Publishers Weekly**; July 24, 1987; 232(4): 163.

Borchardt, D. H. *An Australian information policy.* **Australian Academic and Research Libraries**; March 1980; 11(1): 26-40.

Bourne, Charles P. *On-line systems: history, technology, and economics.* **Journal of the American Society for Information Science**; May 1980: 155-160.

Bourne, Charles P. *The world's technical journal literature: an estimate of volume, origin, language, field, indexing and abstracting.* **American Documentation**; 1962; 13(2): 159-68.

Brady, Edward L., Ed. **U.S. Access to Japanese Technical Literature: Electronics and Electrical Engineering.** Washington, D.C.: U.S. Govt. Printing Office; 1986. 155.

Brittain, J.M. **Information and its Users: A Review with Special Reference to the Social Sciences.** Bath: Bath University Press; 1970.

Buckley, Barbara Jo. *The coverage of library/information science periodicals from developing countries by the major abstracting and indexing services.* **IFLA Journal**; 1982; 8(4): 379-387.

Burke, Clare E.; Price, D. de S. *The distribution of* citations *from nation to nation on a field by field basis: A computer calculation of the parameters.* **Scientometrics**; Sept. 1981; 3(5): 363-377.

Byrne, Alex. *How to lose a nation's literature: Database coverage of Australian research.* **Database**; August 1983; 6(3): 10-17.

Cain, Mark E. *Academic and research librarians: Who are we?* **Journal of Academic Librarianship**; 1988; 14(5): 292-296.

Carpenter, Mark P.; Narin, Francis. *The adequacy of the Science Citation Index (SCI) as an indicator of international scientific activity.* **Journal of the American Society for Information Science**; 1981; 32(6): 430-439.

Chan, G.K.L. The foreign language barrier in science and technology. International Library Review; 1976; 8: 317-25.

Christovão, H. T. *The aging of the literature of biomedical sciences in developed and developing countries.* **Scientometrics**; March 1985; 7(3-6): 411-430.

Clement, John. *Sensitive (but unclassified): is a new category in the national interest?* **Bulletin of the American Society for Information Science**; April/May 1987; 13(4): 14-15.

Colson, Harold. *Searching Latin America online: Social sciences and the humanities.* **Database Searcher**; June 1988; 4(6): 20-24.

Conger, Lucinda D. *Parlez-vous online: Language searching on BRS, DIALOG, and ORBIT.* **Database**; August 1984; 7(3): 86-92.

Craig, Daza Fox; Strain, Paula Meise. *Analysis of collection development at the National Library of Medicine.* **Bulletin of the Medical Library Association**; April 1980; 68(2): 197-206.

Cullars, John. *Characteristics of the monograhic scholarship of foreign literary studies by native speakers of English.* **College and Research Libraries**; March 1988; 49(2): 157-170.

Curwen, Peter. **The World Book Industry**. London: Euromonitor Publications; 1986.

Davenport, Lizzie; Cronin, Blaise. *Government policies and the information industry: The balance of interests.* **Aslib Proceedings**; May 1987; 39(5): 159-167.

Davis, Bob. *Federal agencies press data-base firms to curb access to "sensitive" information.* Wall Street Journal; 1987.

de Silva, Fred. *The language of the Oracle: English as a vehicle of dependence.* **Development Dialogue**; 1976: 165-72.

de Solla Price, Derek J. **Little Science, Big Science ... and Beyond**. New York, NY: Columbia University, 1986.

Dickman, John T. *How do secondary services select journals?* **Cleveland Clinic Quarterly**; Winter 1984: 51(4); 575-578.

Dickson, A.J. *Librarians and the foreign language barrier.* **Aslib Proceedings**; Nov. 1979; 31(11): 488-95.

Diodato, Virgil P. *Original language, non-English journals: Weeding them and holding them.* **Science and Technology Libraries**; Spring 1986; 6(3): 55-67.

Ellen, Sandra R. *Survey of foreign language problems facing the research worker.* **Interlending Review**; 1979; 7(2): 31-41.

First Spanish publicly available database. **Online Review**; 1984; 8(3): 209.

Foo-Kune, C.F. *Japanese scientific and technical periodicals: an analysis of their European language content.* **Journal of Documentation**; 1970; 26: 111-19.

Frame, J. Davidson. *National economic resources and the production of research in lesser developed countries.* **Social Studies of Science**; May 1979; 9(2): 233-246.

Frame, J. Davidson; Carpenter, Mark P. *International research collaboration.* **Social Studies of Science**; Nov. 1979; 9(4): 481-97.

Frame, J. Davidson; Narin, Francis. *The international distribution of biomedical publications.* **Federation Proceedings**; May 1977; 36(6): 1790-1795.

Ganley, Oswald H. *Political aspects of communications and information resources in Canada.* **Information Society**; 1981; 1(1): 79-89.

Garfield, Eugene. *English: an international language for science?* Garfield, Eugene. **Essays of an Information Scientist.** Philadelphia: ISI Press; 1977; 1: 19-20.

Garfield, Eugene. *Journal citation studies 25: German journals: what they cite and vice versa.* **Current Contents**; 1976; 19(18): 5-11.

———. *Le Nouveau Défi Américain.* Garfield, Eugene. **Essays of an Information Scientist**. Philadelphia: ISI Press; 1980; 4: 88-94.

Garvey, William D. Communication: The Essence of Science. Oxford: Pergamon; 1979.

Gilliam, Mary Cooper. *Problems in foreign legal acquisitions.* **Law Library Journal**; Winter 1987; 79(1): 125-131.

Goonatilike, Susantha. Aborted Discovery: **Science and Creativity in the Third World**. Totowa, NJ: Biblio Distribution Center; 1984.

Gordon, M. D. *Deficiencies of scientific information access and output in less developed countries.* **Journal of the American Society for Information Science**; 1979; 30(6): 340-2.

Gordon, M. D.; Santman, A. *Language barriers, literature usage and the role of reviews: an international and interdisciplinary study.* **Journal of Information Science**; 1981; 3: 185-89.

Görlach, Manfred. *English as a world language - the state of the art.* **English World-Wide**; 1988; 9(1): 1-32.

Gould, C. J.; Stern, B. T. *Foreign technical literature: a problem of costs, coverage, and comprehension.* **Aslib Proceedings**; Nov. 1971; 23(11): 571-76.

Greenberg, Daniel S. **The Politics of Pure Science**. New York: New American Library, 1967.

Griffith, Belver C. *Science literature--how faulty a mirror of science?* **Aslib Proceedings**; August 1979; 31(8): 381-91.

Gruson, Lindsey. *Speaking in tongues: Motivated by economics, interest in foreign language study increases in the United States.* **Ann Arbor News**; Dec. 3, 1986: B2, Col. 1.

Guha, Bimalendu. **Study on the Language Barrier in the Production, Dissemination and Use of Scientific and Technical Information with Special Reference to the Problems of the Developing Countries.** Paris: General Information Programme and UNISIST, Unesco; Dec. 1985.

Guinjoán, Aurelio Pérez; Büttenklepper, Alfredo; Sandoval, Armando; Almada de Ascencio, Margarita. *Research in Latin America: A bibliometric approach.* **Inforum**; May 1983; (3): 1-12.

Haag, D. E. *Barriers limiting the usefulness of published information in the research environment.* **Special Libraries**; July 1984; 75(3): 214-20.

Hagstrom, Warren. **The Scientific Community**. New York: Basic Books; 1965.

Hamaker, Charles. *Caveat emptor: foreign imprints in the U. S. marketplace - Tips for new selectors.* **RTSD Newsletter**; Fall 1987; 12(4): 48-49.

Hancock, Ian F; Angogo, Rachel. *English language in East Africa.* Bailey, Richard W; Görlach, Manfred. **English as a World Language**. Ann Arbor, MI: Univ. of Michigan Press; 1982: 306-323.

Hanson, C. W. **The Foreign Language Barrier in Science and Technology**. London: ASLIB, 1962.

Hatfield, C. Donald. *Suddenly, the government is defining journalists.* **Tucson Citizen**; 15 October 1988: 7A.

Hayden, R.L. *Toward a national foreign-language policy.* **Journal of Communication**; 1979; 29(2): 93-101.

Herner, S. *American use of Soviet medical research.* **Science**; 4 July 1958; 128(3314): 9-15.

Hilts, Philip J. *Scientists urged to submit work for US review.* **Washington Post**; 8 January 1982: A1, A11.

Hogeweg-de Haart, H. P. **Characteristics of Social Science Information: A Selected Review of the Literature.** Budapest: FID/SD; 1981. (FID Studies in Social Science Information and Documentation, 1).

———. *Social science and the characteristics of social science information and its users.* **International Forum on Information and Documentation**; Jan. 1983; 8(1): 11-15.

Holmstrom, J.E. *The foreign-language barrier.* **Aslib Proceedings**; Dec. 1962; 14(12): 413-25.

Hsieh, Richard K. C. *International Programs of the National Library of Medicine.* **Bulletin of the Medical Library Association**; Jan. 1988; 76(1): 54-57.

Hunter, J. A. *Quantifiable characteristics of bibliographic databases.* **Information Services and Use**; March 1981; 1(1): 31-43.

Huston, Mary M.; Williams, Joe L. *Reseacher response to the politics of information (Perseverance in spite of biases in subject headings).* **Research Strategies**; Spring 1987; 5: 90-93.

Hutchins, W. J. [and others]. **The Language Barrier: A Study in the Depth of the Place of Foreign Language Materials in the Research Activity of an Academic Community.** Sheffield, England: Univ. of Sheffield, Postgraduate School of Librarianship and Information Science; 1971.

Hutchins, W. J.; Pargeter, L. J.; Saunders, W. L. *University research and the language barrier.* **Journal of Librarianship**; Jan. 1971; 3(1): 1-25.

Inhaber, Herbert. *Where scientists publish.* **Social Studies of Science**; August 1977; 7(3): 388-394.

ISDS. **International Serials Data System.** Paris: ISDS International Centre; 1987.

The ISDS Data Base. Statistical Data. **News from ISDS**; 1985; (no. 9): 1-2.

The ISDS Database. Statistical Data. **News from ISDS**; 1987; (no. 18): 1-3.

Jacobs, Gloria. *Indexing and abstracting services' coverage of Soviet English language periodicals.* **Reference Librarian**; Spring 1987; (no. 17): 123-130.

Japanese markets and techologies on-line. **Data Base Informer**; March 1885; 2(3): 3-6.

JCIST: The Japan Information Center of Science and Technology. Tokyo: JCIST; 1985.

Judge, Peter J. *Dependence, interdependence, and independence: the role of the government in the development of information services.* **National Information Policy Seminar**; December 7-8, 1981; Canberra, Australia. Ultimo, New South Wales, Australia: The Library

Association of Australia; 1981.

Kaplan, Robert. *Languages and science policies of new nations.* **Science**; 12 August, 1983; 221: 609-613.

Keller, Edward; Röel, Ronald. *Foreign languages and U. S. cultural policy: An institutional perspective.* **Journal of Communication**; Spring 1979; 29(2).

Kertesz, F. *How to cope with the foreign language problem: Experience gained at a multi-disciplinary laboratory.* **Journal of the American Society for Information Science**; 1974; 25(2): 86-104.

King, Donald W. **Statistical Indicators of Scientific and Technical Communication, 1960-1980.** Washington, DC: Government Printing Office; 1976.

Kopp, James J. *Research and writing in the history of health sciences, 1970-82: A quantitative analysis of NLM's HISTLINE database.* **Bulletin of the Medical Library Association**; April 1985; 73(2): 146-152.

Landau, H. B. *A secondary publisher's views on federal information policy: a viewpoint.* **Government Publications Review**; January/February 1988; 15(1): 1-10.

Large, J. A. **The Foreign-Language Barrier: Problems in Scientific Communication**. London: Andre Deutsch, 1983.

Lewis, Anthony. *Energy chief seeks restrictions on access to nuclear documents.* **Ann Arbor News**; July 8, 1983: A7, Col. 1.

Lindberg, Donald A. B.; Riecken, Henry W. *Future Programs at the National Library of Medicine.* **Bulletin of the**

Medical Library Association; Oct. 1986; 74(4): 344-352.

Line, Maurice; Roberts, Stephen. *The size, growth and composition of social science literature.* **International Social Science Journal**; 1976; 28(1): 122-159.

Lottman, Herbert R. *The book industry abroad.* **The Book Publishing Annual.** 1985 ed. New York: Bowker, 1985.

Louttit, C. M. *The use of foreign languages by psychologists, chemists, and physicists.* **American Journal of Psychology**; 1957; 70: 314-16.

Love, Erika. *The science of medical librarianship: Investing in the future.* **Bulletin of the Medical Library Association**; Oct. 1987; 75(4): 302-309.

Macias Chapula, C. A. *Mexico's national biomedical bibliography: a proposal for the setting-up of a computer-based indexing system.* **Medecine et Informatique (London)**; July-Sept. 1986; 11(3): 259-67.

Maher, John. *The development of English as an international language of medicine.* **Applied Linguistics**; Summer 1986; 7(2): 206-218.

Mahon, F. V. *Perspectives of access to online information by the developing countries.* **Unesco Journal of Information Science, Librarianship and Archives Administration**; Oct.-Dec. 1982; 4(4): 239-44.

Maloney, Clarence. *The real language problem in South Asia.* **Asian Affairs**; May-June 1974; 1(5): 319-331.

Manten, A. A. *Publication of scientific communication is not identical with comunication.* **Scientometrics**; July 1980; 2(4): 303-308.

Maxwell, Robert. *Data retrieval systems and international publishing.* **Aslib Proceedings**; Jan. 1982; 34(1): 38-45.

McDonald, Kim A. *Science panel recommends steps for encouraging open communication of research results.* **Chronicle of Higher Education**; 14 Dec. 1988; 35(16): 5-6.

Meadows, A. J. *The citation characteristics of astronomical research literature.* **Journal of Documentation**; 1967; 23(1): 28-33.

———, Ed. **The Scientific Journal**. London: ASLIB; 1979.

Mendez, A.; Gomez, I. *The Spanish scientific productivity through eight international databases.* **Scientometrics**; September 1986; 10(3-4): 207-219.

Menou, Michel. *Cultural barriers to the international transfer of information.* **Information Processing and Management**; 1983; 19(3): 121-129.

Michel, D. J.; Knodel, L. C. *Use of bibliographic, abstracting, and computer resources in drug information centers.* **American Journal of Hospital Pharmacy**; March 1988; 45(3): 624-5.

Molino, Enzo. *Databases: considerations of relevance to developing countries.* **Unesco Journal of Information Science and Librarianship**; Oct./Dec. 1982; 4(4): 245-52.

Morton, Bruce; Zink, Steven D. *An interview with Roger K. Summit, President, Dialog Information Services.* **Government Publications Review**; March/April 1988; 15: 97-112.

Moynahan, Sharon. *Faster searching of Spanish and Portuguese language materials on OCLC: A proposal.* **RTSD Newsletter**; 1985; 10(8): 97-98.

Mychko-Megrin, A. Y. *Chinese scientific and biomedical literature and its coverage in international bibliographic and abstracting services.* **Chinese Medical Journal**; April 1987; 100(4): 335-7.

National Academy of Science. **Scientific and Technical Communication (SATCOM Report).** Washington, DC: NAS; 1969.

National Library of Medicine. **Building and Organizing The Library's Collection.** Bethesda, MD: National Institutes of Health; 1986. (Long Range Plan Report of Panel 1).

National Library of Medicine. **Locating and Gaining Access to Medical and Scientific Literature.** Bethesda, MD: National Institutes of Health; 1986. (Long Range Plan Report of Panel 1).

National Library of Medicine. **Long Range Plan: Report of the Board of Regents.** Bethesda, MD: National Institutes of Health; 1986.

National Science Board. **Science Indicators: The 1985 Report.** Washington, D.C.: U.S. Govt. Printing Office; 1985.

National Science Foundation. Directorate for Scientific, Technological, and International Affairs. **International Science and Technology Data Update.** Washington, D.C.: National Science Foundation; Jan. 1985.

Nwagha, Georgiana K. N. *Barriers to the accessibility of scientific journals in Nigeria.* **International Library Review**; 1980; 12: 201-207.

Oliver, Cl. *Les français sont-ils mal-aimés de la littérature médicale anglo-saxonne?* **Journal des Maladies Vasculaires**; 1987; 12: 299.

Orient, Jane M. *A critique of Western medical commentary on the People's Republic of China.* **Archives of Internal Medicine**; Jan. 1984; 144(1): 106-109.

Please read more French. **The Lancet**; June 12, 1982; (no. 8285): 1369.

Pravdic, Nevenka; Aganovic-Boras, Alena; Kritovac, Dubravka. *In search of a "non-citation index" indicator for scientific activity assessment in less developed countries. Case study of Croatia/ Yugoslavia.* **Scientometrics**; July 1988; 14(1-2): 111-125.

Rabkin, Y. M.; Inhaber, H. *Science on the periphery: A citation study of three less developed countries.* **Scientometrics**; March 1979; 1(3): 261-274.

Restivo, Sal P.; Vanderpool, Christopher K., Eds. **Comparative Studies in Science and Society.** Columbus, OH: Charles E. Merrill; 1974.

Richards, Daniel T; Moll, Joy K. *International exchange of scientific literature by U. S. academic health sciences libraries: A literature review and survey of current activities.* **Bulletin of the Medical Library Association**; Oct. 1982; 70(4): 369-373.

Richards, Pamela Spence. *Information science in wartime: Pioneer documentation activities in World War II.* **Journal of the American Society for Information Science**; September 1988; 39(5): 301-306.

Rogers, Michael. *Getting warmer ... the race is on to find materials that superconduct at room temperature.*

Newsweek; July 6, 1987; 110(2); 42.

Rose, Caroline B; Rose, Arnold. **Sociology: The Study of Human Relations**. New York, NY: Knopf; 1969.

Rosenberg, Victor. [Informal accumulated papers for a study addressing languages in databases]. Ann Arbor, MI: 1981.

Rosenberg, Victor. *Information policies of developing countries: the case of Brazil.* **Journal of the American Society for Information Science**; July 1982; 33(4): 203-207.

Rosenberg, Victor; Whitney, Gretchen, Eds. **The Transfer of Scholarly, Scientific and Technical Information between North and South America: Proceedings of a Conference**. Metuchen, NJ: Scarecrow Press, 1986.

Russ, Charles V. J. *The geographical and social variation of English in England and Wales.* Bailey, Richard W; Görlach, Manfred. **English as a World Language**. Ann Arbor, MI: Univ. of Michigan Press; 1982.

Sandoval, Armando M. *The biomedical manuscripts drain from Latin America.* **Unesco Bulletin for Libraries**; January-February 1974; 28(1): 10-16.

Sandoval, Armando M; Pérez-Guinjoán, Aurelio; Franco, Consuelo. *A current Latin American bibliography compiled from non-Latin American journals: a bibliometric study.* **Inforum**; May 1983; (3): 33-39.

Saunders, W. L. *The foreign language and translation problems of the social sciences.* **Aslib Proceedings**; April 1972; 24(4): 233-41.

Sekerak, R. J. *A comparison of journal coverage in Psychological Abstracts and the primary health sciences indexes: Implications for cooperative serials acquisitions and retention.* **Bulletin of the Medical Library Association**; July 1986; 74(3): 231-3.

Sengupta, I. N. *Impact of scientific serials on the advancement of medical knowledge: An objective method of analysis.* **International Library Review**; Oct. 1972; 4(4): 169-95.

Shah, M. A. *Index Medicus. The JNM struggle for inclusion is far from over.* **Journal of Nurse Midwifery**; March-April 1985; 30(2): 69-70.

Shearer, Kenneth D. *The geography of information.* **Geolinguistics**; 1984; 10: 1-11.

——— *The geolinguistics of information: English vis-à-vis other major languages.* **International Library Review**; July 1986; 18(3): 223-230.

——— *How much intellectual capital is banked in the principal human languages?* **International Library Review**; July 1985; 17(3): 235-246.

Smith, Alan G. R., Ed. **Science and Society in the Sixteenth and Seventeenth Centuries.** New York, NY: Science History Publications, 1972.

Smith, Elmer V; Janson, Florentia Scott. *Exporting the American (information) revolution: The international impact of the National Library of Medicine.* **Bulletin of the Medical Library Association**; Oct. 1986; 74(4): 339-343.

Smith, Kent A. *The National Library of Medicine: From MEDLARS to the sesquicentennial and beyond.* **Bulletin of the Medical Library Association**; Oct. 1986; 74(4): 325-332.

Sprehe, J. Timothy. *OMB Circular no. A-130, the management of federal information resources: Its origins and impact.*

Government Information Quarterly; 1987; 4(2): 189-196.

Stankus, Tony; Schlessinger, Rashelle; Schlessinger, Bernard S. *English language trends in German basic science journals: a potential collection tool.* **Science and Technology Libraries**; Spring 1981; 1(3): 55-66.

Starr, S. Frederick. *English dethroned.* **Change**; May 1978; 10(5): 26-31.

Stavik, Anton. *Information and language barriers: Some background and suggestions for action.* **International Forum on Information and Documentation**; July 1987; 12(3): 8-14.

Stefaniak, Barbara. *Use of bibliographic databases for scientometric studies.* **Scientometrics**; September 1987; 12(3-4): 149-161.

Stewart, June L. *The literature of politics: a citation analysis.* **International Library Review**; 1970; 2: 329-53.

Storer, Normal H. *The iternationality of science and the nationality of scientists.* **International Social Science Journal;** 1970; 22(1): 80-93.

Storer, Norman H. The Social System of Science. New York: Holt, Rinehart and Wilson; 1966.

Strategy aims to block Soviet access to U.S. scientific papers. **Ann Arbor News;** Sept. 23, 1982.

Swales, J[ohn] M. *English language papers and author's first language: Preliminary explorations.* **Scientometrics**; July 1985; 8(1-2): 91-101.

Swales, J[ohn] M. *Discourse communities, genres and English as an international language.* **World Englishes**; Summer 1988; 7(2): 211-220.

———. *English as the international language of research.* **RELC Journal**; June 1985; 16(1): 1-7.

———. *ESP in the big world of reprint requests.* **English for Special Purposes**; 1986; 5(1): 81-85.

Swanson, David W.; Swanson, David K. *Paranoid disorder: The Soviet viewpoint.* **American Journal of Psychiatry**; July 1986; 143(7): 825-830.

Tell, Bjorn V. *Information consciousness and knowledge enhancement in the LDCs in view of a New International Information Order.* **International Forum on Information and Documentation**; 1984; 9(1): 3-9.

Thorp, R. G. et al. *The foreign language barrier: A study among pharmaceutical research workers.* **Journal of Information Science**; 1988; 14: 17-24.

Tobey, Ronald C. **The American Ideology of National Science, 1919-1930**. Pittsburgh, PA: Univ. of Pittsburgh Press; 1971.

Todd, Loreto. The English language in West Africa. Bailey, Richard W; Görlach, Manfred. English as a World Language. Ann Arbor, MI: Univ. of Michigan Press; 1982: 281-305.

U. S. Congress. House. Committee on Government Operations. **International Information Flow: Forging a New Framework**. (H. Rept. 96-1535). Washington, DC: Government Printing Office, 1980.

U. S. Congress. House. Committee on Government Operations. **International Data Flow**. Hearings, March 10, 13, 27,

April 21, 1980. Washington, D.C.: GPO, April 1980.

U. S. Congress. House. Committee on Science and Technology. Subcommittee on Science, Research and Technology. **The Availability of Japanese Scientific and Technical Information in the United States.** Hearings, March 6, 7, 1984. Washington, DC: Government Printing Office, 1984.

U. S. Congress. House. Committee on Science and Technology. Subcommittee on Science, Research and Technology. **The Role of Technical Information in U. S. Competitiveness with Japan.** Hearings, June 26, 27, 1985. Washington, DC: Government Printing Office, 1985.

U. S. Congress. House. Committee on Science and Technology. Subcommittee on Science, Research and Technology, and Subcommittee on Natural Resources, Agriculture Research and Environment. **Impact on U. S. Scientific Research of Proposal to Withdraw from Unesco.** Hearings, March 8, 15, 1984. Washington, DC: Government Printing Office, 1984.

Unesco. **Information Policy Objectives (UNISIST Proposals).** Paris: Unesco, 1974.

Unesco. **Scientific and technical translating and other aspects of the language problem.** Paris: Unesco, 1957.

Unesco. **Statistical Yearbook.** Paris: Unesco, 1970-87.

Valdéz, M. Isabel. *Third World countries and conflicting ideologies of the Information Age.* Slack, Jennifer Daryl; Fejes, Fred. **The Ideology of the Information Age.** Norwood, NJ: Ablex, 1987: 200-219.

Vessuri, Hebe M. C. *The social study of science in Latin America.* **Social Studies of Science**; Aug. 1987; 17(3): 519-54.

Vinkler, P. *Weighted impact of publications and relative contribution score. Two new indicators characterizing publication activity of countries.* **Scientometrics**; July 1988; 14(1-2): 161-163.

Watkins, Beverly T. *Teachers urged to take advantage of public's view that languages are important.* **Chronicle of Higher Education**; 7 Dec. 1988; 35(15): 15-16.

Wellisch, Hans. *Linguistic and semantic problems in the use of English-language information services in non-English-speaking countries: Or, how to install an elevator in the tower of Babel.* **International Library Review**; 1973; 5(3): 147-62.

Welsch, Erwin K. *A collection development model for foreign literatures.* **Collection Management**; Spring 1985; 7(1): 1-11.

Williams, Martha E.; Lannom, Laurence. *Lack of standardization of the journal title data element in databases.* **Journal of the American Society for Information Science**; May 1981; 32(3): 229-233.

Wood, D. N. *The foreign language problem facing scientists and technologists in the United Kingdom: report of a recent survey.* **Journal of Documentation**; 1967; 23(2): 117-130.

Woolston, John E. *Information exchange in a North-South context: is there more to gain through cooperation than in trying to establish new markets?* **Aslib Proceedings**; Jan. 1984; 36(1): 7-14.

Zakhariev, Zakhari. *Coexistence Between Written Languages and Socio-Cultural Problems of the Developing Countries.* **Babel**; 1980; 26(1): 23-28.

INDEX